Waste Management and Environmental Impact XI

WIT*PRESS*

WIT Press publishes leading books in Science and Technology.
Visit our website for the current list of titles.
www.witpress.com

WIT*eLibrary*

Home of the Transactions of the Wessex Institute.
Papers contained in this volume are archived in the WIT eLibrary in volume 257 of WIT Transactions on Ecology and the Environment (ISSN 1743-3541).
The WIT eLibrary provides the international scientific community with immediate and permanent access to individual papers presented at WIT conferences.
Visit the WIT eLibrary at www.witpress.com.

ELEVENTH INTERNATIONAL CONFERENCE ON WASTE MANAGEMENT AND ENVIRONMENTAL AND ECONOMIC IMPACT ON SUSTAINABLE DEVELOPMENT

Waste Management and Environmental Impact 2022

Conference Chairmen

Juan Casares
University of Santiago de Compostela, Spain
Member of WIT Board of Directors

Massimiliano Lega
University of Naples Parthenope, Italy

Elena Rada
Insubria University of Varese, Italy

International Scientific Advisory Committee

Organised by

Wessex Institute, UK

Parthenope University of Naples, Italy

Insubria University, Italy

Sponsored by

WIT Transactions on Ecology and the Environment

International Journal of Environmental Impacts

WIT Transactions

Wessex Institute
Ashurst Lodge, Ashurst
Southampton SO40 7AA, UK

We would like to express thanks to all the conference Chairs and members of the International Scientific Advisory Committees for their efforts during the 2022 conference season.

Conference Chairs

International Scientific Advisory Committee Members 2022

Waste Management and Environmental Impact XI

Editors

Juan Casares
University of Santiago de Compostela, Spain
Member of WIT Board of Directors

Massimiliano Lega
University of Naples Parthenope, Italy

Elena Rada
Insubria University of Varese, Italy

WIT*PRESS* Southampton, Boston

Editors:
Juan Casares
University of Santiago de Compostela, Spain
Member of WIT Board of Directors

Massimiliano Lega
University of Naples Parthenope, Italy

Elena Rada
Insubria University of Varese, Italy

Published by

WIT Press
Ashurst Lodge, Ashurst, Southampton, SO40 7AA, UK
Tel: 44 (0) 238 029 3223; Fax: 44 (0) 238 029 2853
E-Mail: witpress@witpress.com
http://www.witpress.com

For USA, Canada and Mexico

Computational Mechanics International Inc
25 Bridge Street, Billerica, MA 01821, USA
Tel: 978 667 5841; Fax: 978 667 7582
E-Mail: infousa@witpress.com
http://www.witpress.com

British Library Cataloguing-in-Publication Data

A Catalogue record for this book is available
from the British Library

ISBN: 978-1-78466-463-3
eISBN: 978-1-78466-464-0
ISSN: 1746-448X (print)
ISSN: 1743-3541 (on-line)

The texts of the papers in this volume were set individually by the authors or under their supervision. Only minor corrections to the text may have been carried out by the publisher.

Preface

This year as a collateral impact of the COVID-19 pandemic, WIT has combined two of their most significant Conferences into one. The Conference, on-line, titled Waste Management and Environmental Impact allowed us to gauge how things are evolving in environmental matters specially those regarding waste and impact.

Just one single case: we have lived in the past months with those horrendous videos of a large mass of plastic waste surrounding Easter Island, and similar places, in the northwest of the Pacific Ocean. The world has produced more plastics since 2000 than all the previous decades together. At the present pace, in 2050, there will be more tons of plastic waste in the oceans than the weight of fish living there.

And we shouldn't be surprised that all planet systems are involved in an exponential mode of behaviour. The Meadows Report, published fifty years ago, warned us all about the cause–effect relationship associated with our way of life in developed countries, and the need to find an equilibrium. Nothing can go on forever in a finite planet but we, in our arrogance, disagreed, now only to find that we have overshot at least three of the physical limits (carrying capacity) of the planet. Moreover, it seems that more are on the way, and sooner rather than later.

Papers presented at the Conference on Environmental strategies and tactics; Waste monitoring and management; Air soil and water pollution; and Circular Economy, highlighted to all attendants and interested readers the cases mentioned in various environments. There was a good mix of attendees, from young scientists from emergent countries through to 'old timers' from more developed regions, with presentations on the use of water, the oil of the present century, and solid, hazardous, and non-hazardous waste receiving a great deal of attention.

We, at WIT are aware that this current time, since early 2020, has been extremely difficult for all, but especially for the research community to bring forward their findings. That is why we have organized extra activities like the webinars in the months following the conferences to enable the delegates to keep in touch, to make you all feel that we were next to you, even if we could not do it in person as we have been doing over the last four decades. Your support is greatly appreciated. And so it is that we have decided to include the accepted papers in a volume of the WIT Transactions on Ecology and the Environment, indexed in Scopus.

The Editors wish to thank the Organizing Committee, who edited the proceedings, and the members of the International Scientific Advisory Committee who have peer-reviewed the submitted papers. Special thanks go to the Conference Coordinator Ms Marta Graczyk, Mr Alan Morgan, the person that made the on-line process run so smoothly, and the production team of WIT Press, to finally get this book on the shelves.

The Editors, 2022

Contents

Section 3: Strategies for waste management: EU and non-EU case studies

Section 4: Waste monitoring and management

Section 5: Environmental strategies and tactics

SECTION 1
AIR, SOIL AND WATER POLLUTION

HIERARCHICAL MONITORING OF WATER QUALITY: COORDINATING THE SPATIOTEMPORAL RESOLUTION OF MULTILAYER AND MULTISPECTRAL SENSORS TO CHARACTERIZE POLLUTION

GABRIELE MEDIO[1], VINCENZO SEVERINO[1], ROBERTA TETA[2], THEODORE ENDRENY[3] & MASSIMILIANO LEGA[1]
[1]Department of Engineering, University of Naples Parthenope, Italy
[2]Department of Pharmacy, University of Naples Federico II, Italy
[3]Department of Environmental Resources Engineering, State University of New York College of Environmental Science and Forestry, USA

ABSTRACT

Pollutant discharges into the environment can have significant adverse impacts on human health and biodiversity, and the detection of these discharges is hindered by the large spatial and temporal scales across which illicit activities also take place. To assist in environmental monitoring of receiving waters this research developed a hierarchical monitoring program to coordinate the spatiotemporal resolution of multilayer and multispectral sensors. The hierarchical monitoring refers to sensors carried by platforms located at different altitudes, operating at different spatial, temporal and spectral resolutions. The unique combinations of sensors generate information layers characterized by different relationships between several parameters of targeted area, making it possible to focus on specific aspects of the phenomena and to produce novel and complementary results. The proximal sensing instruments, such as drones which can operate closest to the target, often have the highest spatial and temporal resolution, and sufficient multispectral capacity, to provide a link or bridge between in-situ sampling and higher altitude remote sensing. The hierarchical monitoring study was conducted on the Domitia coast in southern Italy, where the Regi Lagni channel discharges. The Regi Lagni is contaminated by discharges from five wastewater plants, and receives illicit discharges in other sections, notably a wide wetland area. The multispectral and multilayer analysis evaluated the impacted area, using multispectral output of the proximal sensing acquisition drone system as a bridge layer to contextualize information coming from satellite acquisition. This operation was able to quantify the spatial impact of pollutant discharges through a classification process related to the anomalies of the multispectral response of coastal waters, using algorithms based on spatial variation of pollutant concentrations. This study, extended over statistically significant time intervals, can provide useful information for the decision-making/technical processes related to the classification and quality control of coastal waters.

Keywords: environmental impact assessment, pollution, coastal waters, multispectral analysis, multilayer analysis, proximal sensing, remote sensing, hierarchical monitoring.

1 INTRODUCTION

The marine coastal zone, like the heart of animals, regulates the circulation of critical ecological, economic and social processes; however, anthropogenic stress poses a serious threat to the coastal environment. Among the phenomena of greatest concern is the pollution of coastal waters via waterways, such as rivers, canals and direct discharges. The substances released in this type of pollution can include: organic matter, synthetic organic compounds (e.g. PCBs and pesticides such as DDT and residues), microorganisms, nutrients (mainly nitrogen and phosphorus), petroleum, waste, heavy metals, radionuclides, and so on. Given the complexity and on-going dynamic evolution of these coastal environments, appropriate coastal management requires data collection whose nature and frequency of space–time acquisition are suitable for the description and the evaluation of the phenomenon of interest.

Conventional in-situ, low-frequency detection methods are often inadequate to capture the complexity and dynamism of these environments. Advances in environmental monitoring provide suitable techniques and technologies for a more complete description of the investigated phenomenon along coastal zones. Using a combination of remote and proximal sensing allows for hierarchical monitoring [1], [2]. By using sensors transported on platforms located at different altitudes hierarchical monitoring produces information layers characterized by different relationships between the spatial resolution of the image and the size of the area surveyed. Consequently, with the various platforms there is a tendency to focus on different aspects and to produce complementary results [3]–[6].

Recent research utilizing hierarchical monitoring together with bioindicators [7], [8] has emphasized the importance of the proximal sensing information in the study of coastal phenomenon related to algal blooms. The information obtained with proximal sensing are closest to the ground and have high spatial and temporal resolutions, which are generally comparable with the data obtained on the ground by conventional in-situ sampling of water quality concerns such as algal blooms. This proximity allows the information regarding bioindicators derived from remote sensing to be validated with greater certainty [9]–[11], enabling management to respond to potentially harmful exposure to algal blooms.

This paper presents a real application of proximal sensing within hierarchical monitoring with bioindicators as an innovative technology for coastal management and environmental impacts during COVID, September 2021. The study site is of Regi Lagni channel on the coastal area of the Domizia coast, located in southern Italy in the Campania region (Fig. 1). Regi Lagni are essentially a network of canals created to hydraulically reclaim an area that extends for 1095 km^2 in the municipalities of Naples, Caserta, Avellino and Benevento. The canalization and hydraulic reclamation works started in Roman times, and then underwent a major hydraulic reorganization during the Spanish Viceroyalty in the early 1600s. The 1973 cholera infection motivated the start of the sanitary reorganization of the area, so many of the drainage channels were used for the collection and treatment processes of sewage waste. In particular, five large purifiers were built whose treated water discharges directly into the channel of Regi Lagni. These works were supposed to solve the sanitation problem of wastewater and improve the quality of coastal waters.

Several environmental studies, including that of ENEA (Italian National Agency for New Technologies, Energy and Sustainable Economic Development), have highlighted various problems concerning the wastewater collection and treatment system and the impact on this Regi Lagni waterway [12]. Furthermore, several police investigations have revealed the direct release of toxic waste (e.g. from companies) and illegal wastewater discharge (e.g. from residential areas, dairy farms and livestock farms) into Regi Lagni [13]. Consequently, for several years, the segment of coast near the mouth of Regi Lagni, located in the Municipality of Castel Volturno, has been one of the most polluted areas of the Domizia coast. The large amount of pollutant discharged into the coastal environment can have significant adverse impacts on human health and biodiversity, and the detection of these discharges is hindered by the large spatial and temporal scales across which illicit activities also take place.

The aim of this study is to reveal hierarchical monitoring methodologies that assist with environmental risk assessment of receiving waters, where monitoring coordinates the spatiotemporal resolution of multilayer and multispectral sensors. The multispectral and multilayer analysis evaluated the impacted area with use of multispectral output from proximal sensing acquisition by a UAV (unmanned aerial vehicle), otherwise known as a drone system. The UAVs are flying platforms characterized by the absence of the pilot on

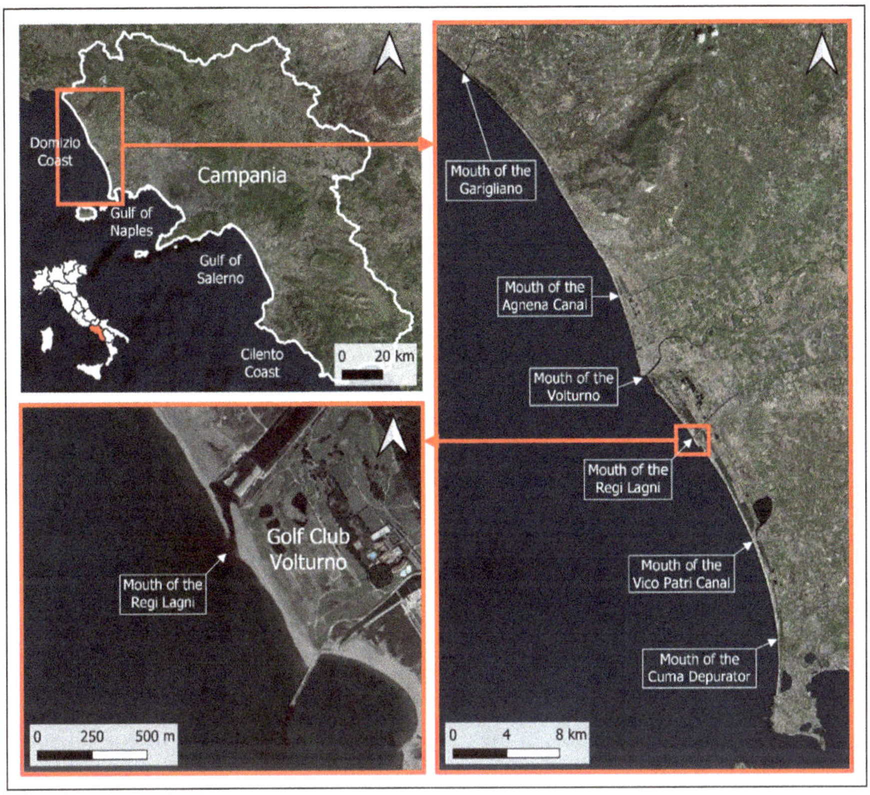

Figure 1: Study sites are to the west of the Italian Gulf of Naples (upper left), along the Domitian Coast (right), and focused about coastal outlet of the Regi Lagni drainage (lower left).

board. The drone system was used as a bridge layer in the multi-layer monitoring to contextualize information coming from satellite acquisition. This study was able to quantify the spatial impact of coastal pollutant discharges through a bioindicator classification process related to the anomalies of the multispectral response of coastal waters, using algorithms based on spatial variation of pollutant concentrations.

2 MATERIALS AND METHODS

2.1 Techniques and technologies

2.1.1 Spectral indices

In the process of remote sensing, the chance of acquiring different bands of the electromagnetic spectrum has allowed the introduction of different spectral indices, which are defined as a simple combination of the reflectance of the spectral bands; spectral indices are generally used to emphasize some elements of the observed scene, which is extremely useful to identify and study "hidden" and complex phenomena. Concerning this study, it was considered appropriate to use two spectral indices, the Normalized Difference Vegetation Index (NDVI) and the Normalized Difference Turbidity Index (NDTI), explained below.

The NDVI, used for the first time by Rouse et al. [14], allows to identify the vegetation areas and their vigour (linked to the photosynthetic activity of the plant). NDVI can also provide useful information on possible pollution phenomena, in fact, this index is sensitive to the presence of algal blooms and bacteria (e.g. cyanobacteria), inside coastal waters, which are generally triggered by eutrophic conditions due to supply of nutrients contained in waste or contaminated water. The NDVI combines the reflectance of the red (RED) and near infrared (NIR) spectral bands and is defined with the following expression:

$$NDVI = (R_{NIR} - R_{RED})/(R_{NIR} + R_{RED}). \quad (1)$$

The NDTI provides information on the turbidity of the water. Because the turbidity of the water is due to the dispersion and absorption of light caused by the presence of suspended particles, high values of the NDTI indicate cloudy water, while low values indicate clear water. Generally, this index tends to emphasize the presence of plumes that water courses form when they meet coastal waters; the greater turbidity of the inlet waters is generally due to suspended particles, such as clay, silt, microorganisms and various substances, deriving from water courses. The NDTI index combines the reflectance of the red (RED) and green (GREEN) spectral bands:

$$NDTI = (R_{RED} - R_{GREEN})/(R_{RED} + R_{GREEN}). \quad (2)$$

2.1.2 Platforms and sensors for proximal and remote sensing

In environmental studies, hierarchical monitoring represents an important source of information provided by sensors carried on platforms located at different altitudes. These sensors provide information layers characterized by different ratios between the spatial resolution of the image and the size of the detected area. Consequently, through hierarchical monitoring thanks to the various platforms it is possible to focus on different aspects of a phenomenon and to produce complementary results.

This study applied hierarchical monitoring with two types of platforms, each with specific sensors. The higher layer was monitored by satellite platforms of the Copernicus Sentinel 2 mission. The lower, proximal layer was monitored by the DJI Mavic 2 Pro drone. Through the sensors carried onboard these platforms, remote and proximal information levels to the surface of interest were obtained respectively.

The Copernicus Sentinel 2 mission includes a constellation of two satellites located in the same polar orbit with 180° out of phase distance. These twin satellites (2A and 2B) make it possible to obtain broad spectrum and high-resolution multispectral images using an optical instrument that samples 13 spectral bands (see Table 1). The Sentinel 2 products included a collection of orthophotos, each spanning 100x100 km^2 in UTM/WGS84 projection, containing the 13 spectral bands previously listed. In particular, only bands 3, 4 and 8 were used for the calculation of the indices of interest, which makes it possible to obtain elaborations with a pixel spatial resolution of 10 m.

The DJI Mavic 2 Pro drone was equipped with an RGB camera fully stabilized by a three-axis gimbal, with a 1-inch CMOS sensor (developed in collaboration with Hasselblad) to record videos in 4K and to acquire 20-megapixel pictures. In addition to the sensor equipment provided by the Mavic 2 Pro manufacturer, the drone was equipped with a multispectral camera Survey 3 with wide-angle lens (Survey3W) produced by MAPIR. The Survey3 used a Sony Exmor R IMX117 12 MP (Bayer RGB) sensor that was customized with an OCN (orange, cyan, near-infrared) filter to allow detection of the spectral bands of orange, cyan and near infrared (Fig. 2). To use the measurements acquired by the Survey3, they were

Table 1: Spectral bands for the SENTINEL-2 sensors (S2A and S2B) showing spatial resolution (m), central wavelength (nm), bandwidth (nm) and a description, where VNIR is very near infrared and SWIR is shortwave infrared. *(Source: Sentinel-2 MSI Technical Guide – Sentinel Online – esa.int.)*

Band number	Spatial resolution (m)	Central wavelength (nm)		Bandwidth (nm)		Description
		S2A	S2B	S2A	S2B	
1	60	442.7	442.3	21	21	Ultra blue
2	10	492.4	492.1	66	66	Blue
3	10	559.8	559.0	36	36	Green
4	10	664.6	665.0	31	31	Red
5	20	704.1	703.8	15	16	VNIR
6	20	740.5	739.1	15	15	VNIR
7	20	782.8	779.7	20	20	VNIR
8	10	832.8	833.0	106	106	VNIR
8a	20	864.7	864.0	21	22	VNIR
9	60	945.1	943.2	20	21	SWIR
10	60	1,373.5	1,376.9	31	30	SWIR
11	20	1,613.7	1,610.4	91	94	SWIR
12	20	2,202.4	2,185.7	175	185	SWIR

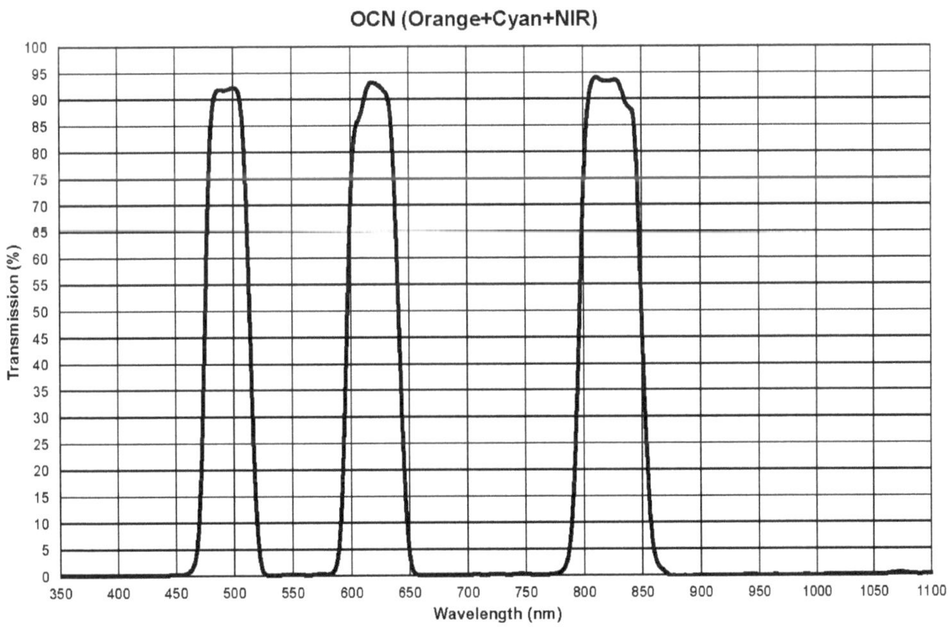

Figure 2: Characteristics of the spectral bands transmitted by the OCN (orange, cyan, near-infrared) filter on the Survey3 sensor carried by the proximal sensing DJI Mavic 2 Pro drone within this mission.

calibrated using a specialized software coupled with a calibrated reflectance target; the latter consists of four surfaces, each of which has known reflectance curves. Prior to an acquisition mission, the target must also be acquired in the same light conditions, taking care not to cover the targets with shadows or debris.

The proximal acquisition phases of the study used two cameras: RGB (red, green, blue) and multispectral combination with detection of six spectral bands. The NDTI index was calculated with the RGB camera bands, while the NDVI index was calculated with the multispectral camera bands. The spatial resolution of the proximal acquisitions used in this study, relative to a maximum altitude of about 350 m, is of the order of 10 cm.

2.1.3 Software

The Sentinel 2 mission orthophotos relating to the area of interest were selected and downloaded from the ESA Copernicus Open Access Hub. For processing of the Survey3 images, two software programs were used in this study. Pre-processing was performed with the Mapir Camera Control (MCC) where this program allowed conversion of RAW + JPG files into TIFF along with calibration of the images. In-depth processing was performed with QGIS software, which enabled calculation of indices using the raster calculator. QGIS allowed customized classes of index values and the false colour scale, as well as a whole series of features that help to visually enhance, through thematic maps, certain aspects of a phenomenon.

2.2 Procedure

The study was conducted following a top-down approach, that is, using first remote platforms and then those proximal to the surface of interest; this strategy allowed to initially have a global view of the anomalies of the multispectral response of the coastal waters of a stretch of the Domizia coastline; after that, using the Mavic 2 Pro, it was decided to accurately analyse the mouth of Regi Lagni, located in the territory of Castel Volturno (Caserta). To optimize the use of the drone, the date of the flight mission was chosen to make it coincident with one of the dates of passage of the Sentinel 2 satellites at the place of interest; at the same time, among the various available dates, the one that presented favourable weather conditions was chosen; in essence, the day that satisfied these conditions was 24 September 2021. A flight plan was programmed through the DJI Ground Station Pro (DJI GS PRO) application, and before take-off a series of checks were carried out. The mission was complemented by a series of supporting information channels, including the weather conditions obtained from the Grazzanise airport meteorological station (Table 2).

Table 2: Weather data provided by the Meteorological Station of Grazzanise Airport (CE) in the southern coast of Italy.

Medium temperature	Minimum temperature	Maximum temperature	Dew point	Average humidity
20°C	13°C	27°C	13°C	68%
Minimum humidity	Maximum humidity	Average pressure above sea level	Phenomena	Weather conditions
34%	100%	1020 mb	Nothing	Clear

Figure 3: Mouth of Regi Lagni drainage entering the Italian coastal area, as seen from the DJI MAVIC 2 Pro drone with the RGB (red, green, blue) image.

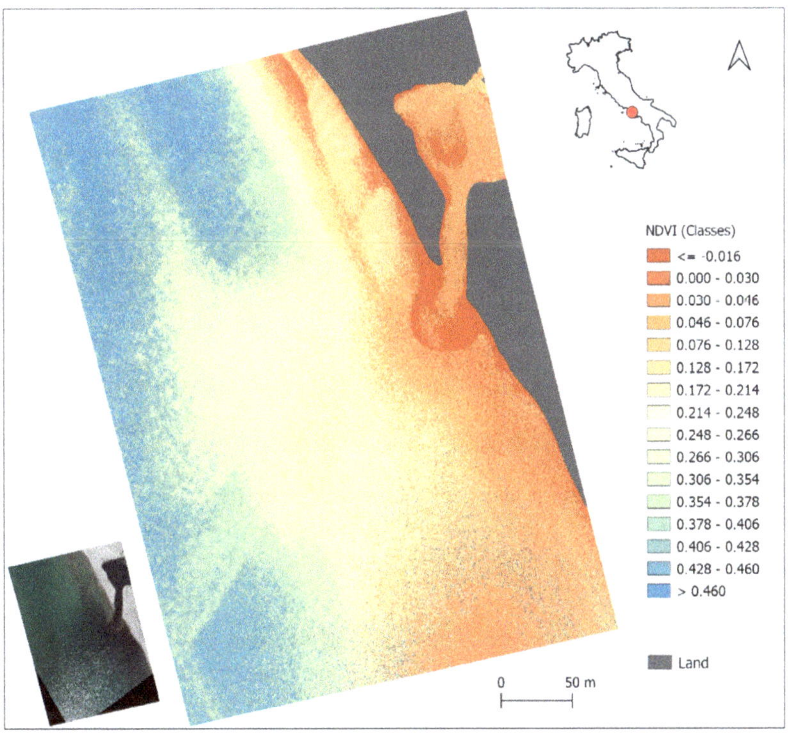

Figure 4: Mouth of Regi Lagni drainage entering the Italian coastal waters, shown in colour with the modified normalized difference vegetation index (NDVI) of the image acquired by DJI Mavic 2 Pro drone Survey3 sensor.

3 RESULTS AND DISCUSSION

The pollution plume discharging into the coastal waters from the Regi Lagni drainage was first sighted through the RGB image captured with the 20 Megapixel camera of the drone (Fig. 3). Even without the use of any spectral index, the Regi Lagni canal waters discharging at the coast were clearly distinguishable from that of the sea, due to the different colour, which was the first clue of the pollutant plume.

The initial visible, e.g., RGB, proximal imagery of the plume (Fig. 3) alone would suggest the impact was spatially limited in extent, however its plume extent was larger. Processing the image obtained from the multispectral camera with the NDVI index shed light on the real impacted zone, a much larger area (Fig. 4) than the one detected in the field of visible. Once the plume was characterized at its full extent with NDVI, the monitoring could consider a potential diffusion of pollutants to a much larger area than characterized by the RGB image.

By applying the NDTI index to the image obtained with the drone camera, it was possible to observe a different, yet also very extensive plume (Fig. 5). In particular, the NDTI in this case provided a very detailed delimitation of the possible zones of the plume.

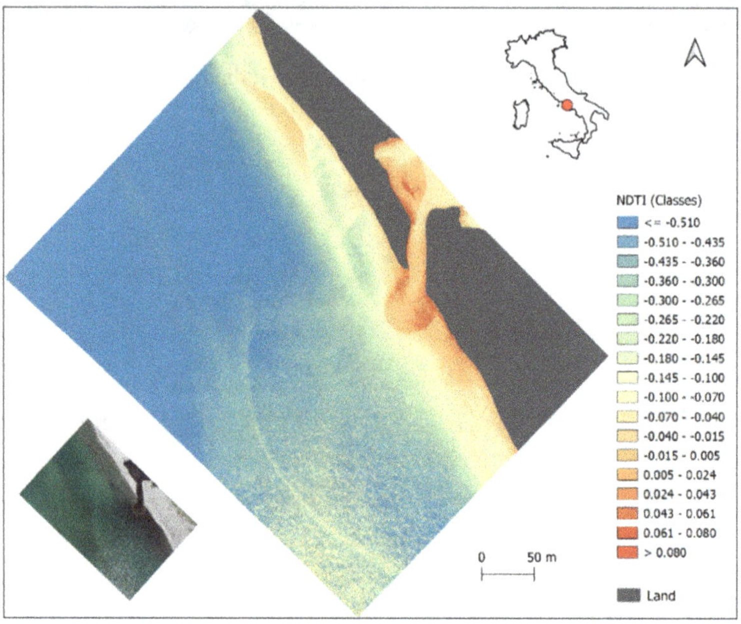

Figure 5: Mouth of the Regi Lagni drainage entering the Italian coastal waters, shown in colour with the modified normalized difference turbidity index (NDTI) of the image acquired by the DJI Mavic 2 Pro drone Survey3 sensor.

Satellite multispectral imagery for 24 September 2021 was then used to elaborate on the spatial extent of the pollution plume, which the drone was unable to fully capture given its proximal distance from the target. The satellite acquisition was processed into the NDTI and provided a spatially broader but coarser 10 m pixel resolution view of the plume (Fig. 6). Data fusion by overlap was then used to merge the two hierarchical monitoring levels, combining the 10 m resolution satellite NDTI with the cm resolution drone NDTI, to obtain enhanced information (Fig. 7).

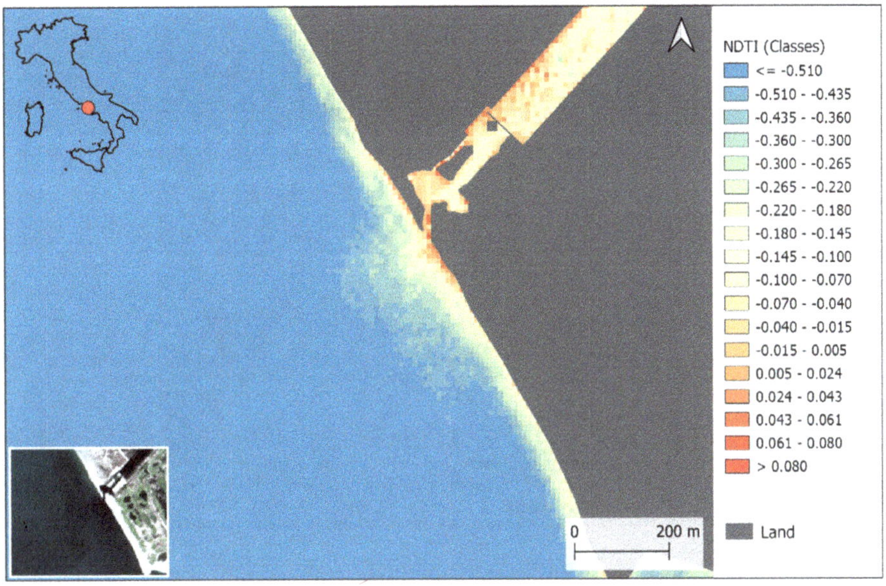

Figure 6: Mouth of the Regi Lagni drainage entering the Italian coastal waters, shown in colour with the modified normalized difference turbidity index (NDTI) of the image acquired by the Sentinel 2 mission. Scale used for comparison.

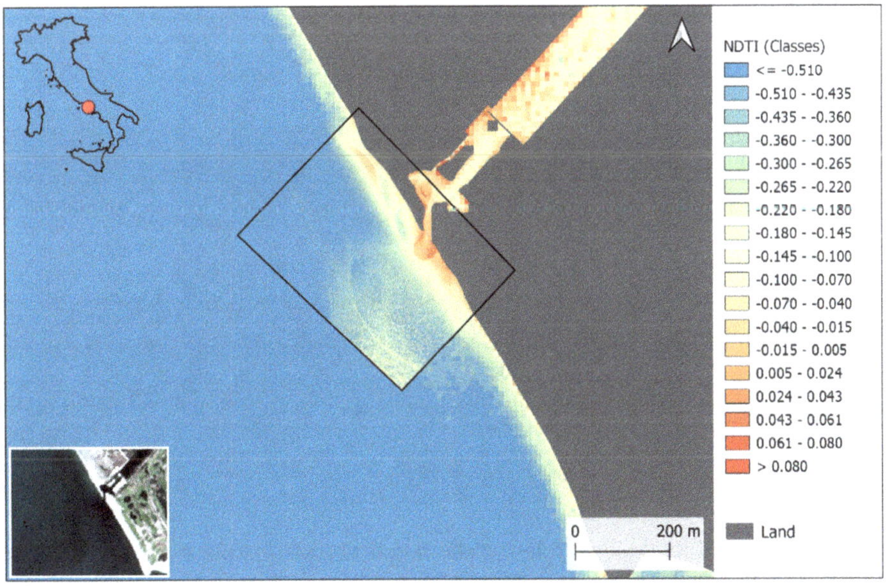

Figure 7: Mouth of the Regi Lagni drainage entering the Italian coastal waters, shown in colour with the modified normalized difference turbidity index (NDTI) data fusion of the satellite Sentinel 2 mission and DJI Mavic 2 Pro drone mission.

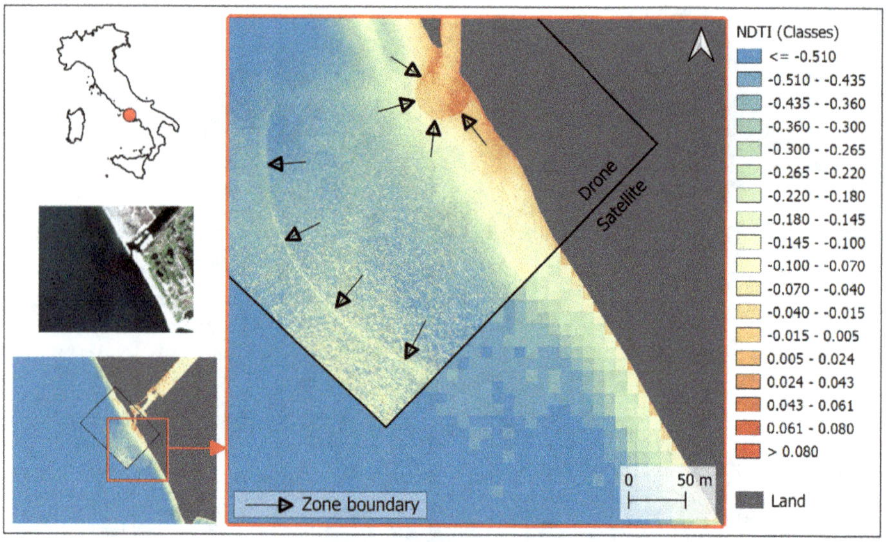

Figure 8: Mouth of the Regi Lagni drainage entering the Italian coastal waters with delimitation of the hierarchical monitoring zones by satellite and drone.

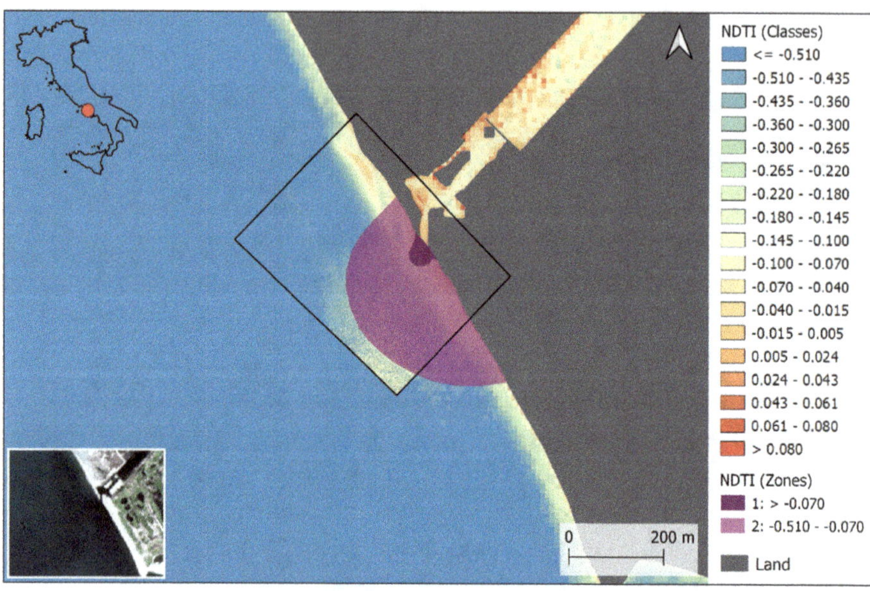

Figure 9: Mouth of Regi Lagni drainage entering the Italian coastal waters with geometric characterization of two zones of the plume based on normalized difference turbidity index (NDTI) with data from the drone.

The data fusion overlap emphasized the substantial differences in spatial resolution between the products of different information levels, and allowed interpretation of the relationships between hierarchical monitoring. From the high spatial resolution drone image it was possible to accurately identify at least two zones of the plume, but the lower spatial resolution satellite image of the outermost plume along the coast cannot be fully defined (Fig. 8).

Geometric processing was used to clarify the intensity of the NDTI values, as sections of the plume and concentration of pollutant. The proximal sensing allowed for a limited spatial extent of two NDTI zones near the drainage outlet to the coast (see Fig. 9). The data fusion between drone and satellite sensing allowed for greater spatial extent and characterization of four NDTI zones (see Fig. 10).

The characterization of the plume zones based on NDTI, with two from the drone (Fig. 9) and four with the data fusion of drone and satellite (Fig. 10) relied on the coordination of hierarchical monitoring. The reconstruction of these two additional zones of the plume by use of satellite data was characterized by less spatial precision and accuracy. This reconstruction performed in QGIS allowed, through polygonal shapefiles, to quantify the extent of the plume surface in square meters (Table 3).

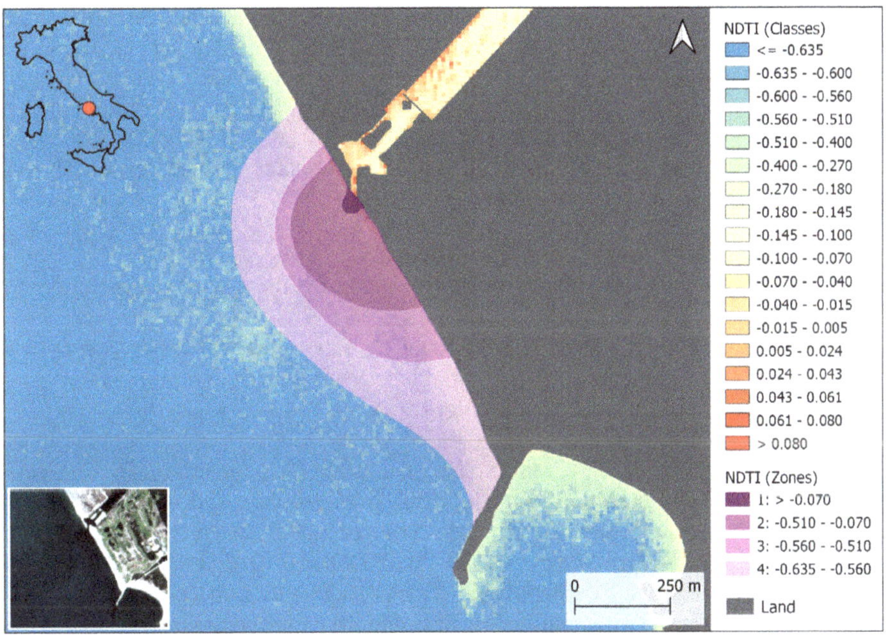

Figure 10: Mouth of Regi Lagni drainage entering the Italian coastal waters with geometric characterization of four zones of the plume based on normalized difference turbidity index (NDTI) with data from the drone and satellite.

Table 3: Areal extension of the plume extent (m^2) in the Italian coastal waters near Regi Lagni, as calculated through the QGIS software analysis.

Extension of the plume surface (m^2)				
Zone 1	Zone 2	Zone 3	Zone 4	total
2087	66774	53360	155329	277550

4 CONCLUSIONS

This study reports how a hierarchical monitoring with specialized sensors and bioindicator indices characterized the pollutant plume for a large, dynamic coastal area. This approach facilitated the focus on specific aspects of the pollution phenomena, and enabled novel and complementary results. Specifically, through this approach it was possible to (a) detect and identify the plume impacting the coastal area near the Regi Lagni drainage, and (b) to define the plume extent and quantify its area in different concentrations of potential pollution.

The proximal sensing instruments, such as drones which can operate closest to the target, often have the highest spatial and temporal resolution, and sufficient multispectral capacity, to provide a link or bridge between in-situ sampling and higher altitude remote sensing. This paper demonstrates how multidisciplinary analysis with coordinated use of remote and proximal sensing was able to quantify the spatial impact of pollutant discharges through a classification process related to the anomalies of the multispectral response of coastal waters, using algorithms based on spatial variation of pollutant concentrations.

This study, when extended over statistically significant time intervals, can provide useful information for the decision-making and technical processes related to the classification and quality control of coastal waters.

ACKNOWLEDGEMENT

This project was supported in part by the University of Naples "Parthenope" (Italy) under "Bando di sostegno alla ricerca individuale".

REFERENCES

[1] Errico, A. et al., SAR/multispectral image fusion for the detection of environmental hazards with a GIS, 2014. DOI: 10.1117/12.2066476.

[2] Lega, M., Casazza, M., Teta, R. & Zappa, C.J., Environmental impact assessment: A multilevel, multi-parametric framework for coastal waters. *International Journal of Sustainable Development and Planning*, **13**(8), pp. 1041–1049, 2018. DOI: 10.2495/SDP-V13-N8-1041-1049.

[3] Lega, M., Kosmatka, J., Casazza, M., De Maio, L., Severino, D. & Teta, R., Environmental monitoring and assessment: A multi-scale and multiparameter case study in Campania Region (S Italy) on the recreational seawater quality evaluation. *Regional Studies in Marine Science*, **39**, 2020. DOI: 10.1016/j.rsma.2020.101465.

[4] Lega, M. & Endreny, T., Quantifying the environmental impact of pollutant plumes from coastal rivers with remote sensing and river basin modelling. *International Journal of Sustainable Development and Planning*, **11**(5), pp. 651–662, 2016. DOI: 10.2495/SDP-V11-N5-651-662.

[5] Di Fiore, V., Cavuoto, G., Punzo, M., Tarallo, D., Casazza, M., Guarriello, S.M. & Lega, M., Integrated hierarchical geo-environmental survey strategy applied to the detection and investigation of an illegal landfill: A case study in the Campania Region (southern Italy). *Forensic Science International*, **279**, pp. 96–105, 2017. DOI: 10.1016/j.forsciint.2017.08.016.

[6] Ferrara, C., Lega, M., Fusco, G., Bishop, P. & Endreny, T., Characterization of terrestrial discharges into coastal waters with thermal imagery from a hierarchical monitoring program. *Water (Switzerland)*, **9**(7), 2017. DOI: 10.3390/w9070500.

[7] Teta, R., Esposito, G., Casazza, M., Zappa, C.J., Endreny, T.A., Mangoni, A., Costantino, V. & Lega, M., Bioindicators as a tool in environmental impact assessment: Cyanobacteria as a sentinel of pollution. *International Journal of Sustainable Development and Planning*, **14**(1), pp. 1–8, 2019. DOI: 10.2495/SDP-V14-N1-1-8.

[8] Esposito, G., Teta, R., Marrone, R., De Sterlich, C., Casazza, M., Anastasio, A., Lega, M. & Costantino, V., A fast detection strategy for cyanobacterial blooms and associated cyanotoxins (FDSCC) reveals the occurrence of Lyngbyatoxin A in Campania (south Italy). *Chemosphere*, **225**, pp. 342–351, 2019. DOI: 10.1016/j.chemosphere.2019.02.201.

[9] Teta, R., Esposito, G., De Sterlich, C., Lega, M. & Costantino, V., Early detection of cyanobacterial blooms and associated cyanotoxins using fast detection strategy. *Journal of Visualized Experiments*, **168**, 2021. DOI: 10.3791/61889.

[10] Teta, R., Sala, G.D., Esposito, G., Stornaiuolo, M., Scarpato, S., Casazza, M., Anastasio, A., Lega, M. & Costantino, V., Monitoring cyanobacterial blooms during the COVID-19 pandemic in Campania, Italy: The case of Lake Avernus. *Toxins*, **13**(7), 2021. DOI: 10.3390/toxins13070471.

[11] Teta, R., Romano, V., Della Sala, G., Picchio, S., De Sterlich, C., Mangoni, A., Di Tullio, G., Costantino, V. & Lega, M., Cyanobacteria as indicators of water quality in Campania coasts, Italy: A monitoring strategy combining remote/proximal sensing and in situ data. *Environmental Research Letters*, **12**(2), 024001, 2017. DOI: 10.1088/1748-9326/aa5649.

[12] http://www.bologna.enea.it/ambtd/regi-lagni/Regi_Lagni_up-to-date_2010/.

[13] Persechino, G., Lega, M., Romano, G., Gargiulo, F. & Cicala, L., IDES Project: An advanced tool to investigate illegal dumping. *WIT Transactions on Ecology and the Environment*, vol. 173, WIT Press: Southampton and Boston, pp. 603–614, 2013. DOI: 10.2495/SDP130501.

[14] Rouse, J., Haas, R., Schell, J. & Deering, D., Monitoring vegetation systems in the great plains with ERTS. *Third ERTS Symposium, NASA*, vol. 1. pp. 309–317, 1973.

POLLUTION OF THE VAAL RIVER SYSTEM IN SOUTH AFRICA: A CASE STUDY

ELKINGTON SIBUSISO MNGUNI
Trans-Caledon Tunnel Authority, South Africa

ABSTRACT

The Vaal River is one of South Africa's strongest flowing rivers. It is home to the Vaal Dam, which supplies water to the Gauteng Province, South Africa's economic hub. Over the years, the failing wastewater treatment system in the catchment area has led to continued pollution of the river, compromising this critical resource and the economy of the region. In 2018, following a public outcry, the South African Human Rights Commission instituted an inquiry into the state of affairs of the region's water and sanitation management issues as well as the level and extent of the pollution problem. The Commission discovered raw sewerage in a stream flowing through the Emfuleni Golf Course, burst sewer pipes on the banks of the Rietspruit, dysfunctional components in the Rietspruit Wastewater Treatment Works, blocked manholes, and children from a nearby school swimming in and consuming polluted water. Raw sewage was found to be discharging from dysfunctional wastewater treatment plants into the receiving Vaal River. In the findings of the report released in February 2021, the Commission found that the Vaal River was polluted beyond acceptable levels. This case study reviews the state of the Vaal River pollution, including the measures that have been designed to address the problem in the short, medium, and long term. It also looks into the successes and challenges being encountered in addressing the Vaal River water pollution problem.

Keywords: Vaal River, pollution, wastewater treatment, interventions.

1 INTRODUCTION

This case study provides the historical background and development of the water pollution crisis in the Vaal River System in the Gauteng Province, South Africa. It starts from the assessment of wastewater treatment capacities conducted in 2008 to a full-blown water pollution crisis due to raw sewage spillages in 2018. It then reviews the intervention efforts that followed a public outcry and an inquiry instituted by the South African Human Rights Commission (the Commission) in 2018, and further action that was taken after the release of its final report in February 2021. The study also provides a brief outline of the progress that has been made until the end of 2021 with concomitant challenges in the implementation of the interventions, as well as making some recommendations that could improve the situation.

2 LOCATION OF THE STUDY AREA

The pollution problem discussed in this study emanates mainly from the Sedibeng District Municipality (DM), which comprises the Lesedi, Midvaal and Emfuleni Local Municipalities, forming the southmost region of the Gauteng Province. The Vaal River forms the southern border of the district municipality. The Sedibeng DM borders with the North-West Province (to the west), Mpumalanga Province (to the east) and the Free State Province (to the south). Fig. 1 shows the location of the study area.

3 PROBLEM DESCRIPTION AND BACKGROUND

The Vaal River is one of South Africa's strongest flowing rivers. The Vaal Dam which supplies water to the Gauteng Province, South Africa's economic hub, is located on the Vaal River which forms the southern border of the province. In their submission to the

www.witpress.com, ISSN 1743-3541 (on-line)
doi:10.2495/WMEI220021

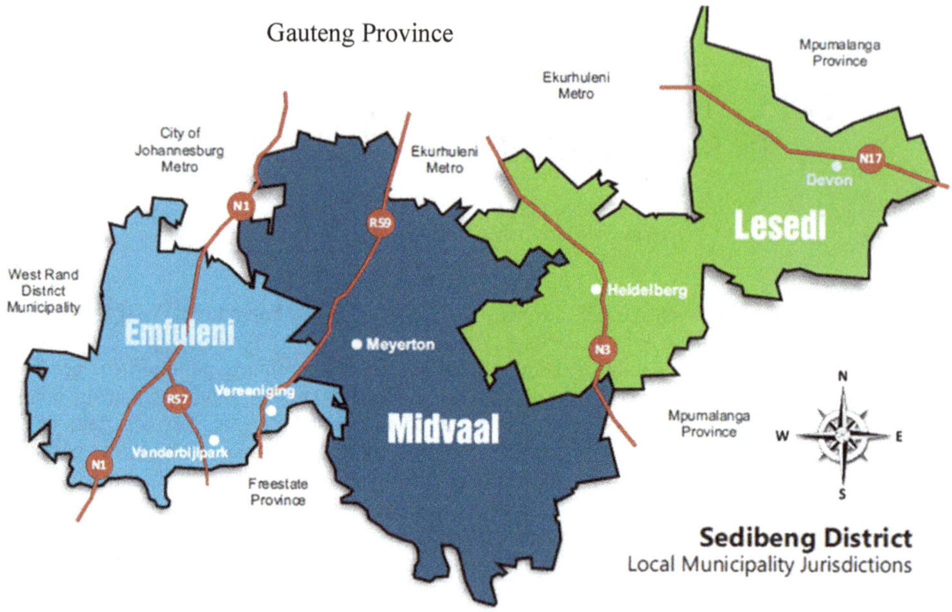

Figure 1: Locality of the study area [1].

Commission, the Water Research Commission, a government entity of South Africa, asserted that the Vaal River was the country's "most economically valuable aquatic ecosystem, and Africa's hardest working river for over a 100 years, and was able, in parts, to recover from mining and industrial pollution" [2].

Over the years, the failing wastewater treatment works in the drainage basins of the river have led to continued pollution of the river, thus compromising this critical resource for social, agricultural, commercial and industrial use, as well as for power generation.

One of the worst affected areas of the province is the Emfuleni Local Municipality (Municipality), a water services authority in terms of South Africa's Water Services Act of 1997 [4]. This is where the majority of the failed wastewater collection networks and wastewater treatment plants are located.

The locations of the three wastewater treatment plants under consideration, namely, the Sebokeng Wastewater Treatment Works (WWTW), the Rietspruit WWTW, and the Leeukuil WWTW, are shown in Fig. 2 which shows the study area. The Municipality is thus the focus of this case study in terms of the river pollution experienced in the region.

3.1 The capacity of wastewater treatment works in 2008

In a study commissioned to investigate wastewater treatment capacity in the Municipality, which was released in October 2008, it was found that the capacity of the wastewater treatment plants in the region would need to be extended by some 275 million litres per day (ML/day) by 2025 [5]. The proposed additional capacity would be composed of 100 ML/day at the extended Sebokeng WWTW, and 175 ML/day at a new WWTW to be constructed within the region. The additional capacity would be required to address the sanitation backlogs at the time and to cater for future growth in the region.

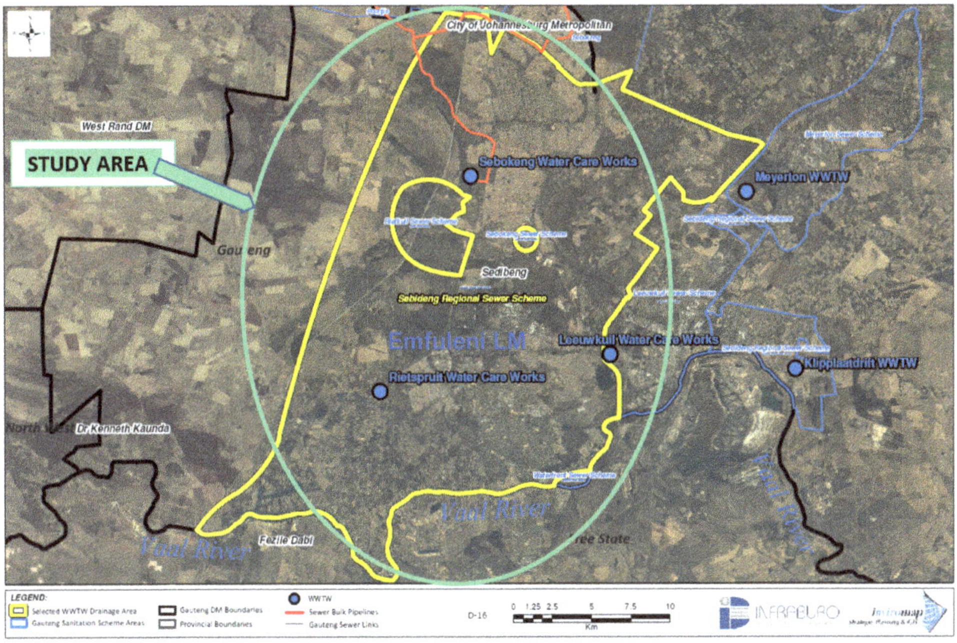

Figure 2: The three WWTWs within the study area [3].

It was also found that some 20 ML/day of surface and ground water infiltrated into the wastewater collector network flowing into the 100 ML/day Sebokeng WWTW. This infiltration problem needed to be addressed in the short term in order to free up much needed hydraulic treatment capacity in that plant.

The new WWTW would need to be built by 2015 in order to meet the treatment capacity requirements for the region by 2025 [5]. The 2008 report led to a proposal, in 2010, to develop and implement a multifaceted project to address the regional wastewater treatment capacity current and future requirements. The projects was referred to as the Sedibeng Regional Sanitation Scheme (SRSS) [3].

3.2 The Gauteng Province's WWTWs Master Plan of 2014

The Department of Water and Sanitation (DWS) further commissioned a study to develop a Master Plan for the Gauteng Province's WWTWs. The master plan was completed and released in April 2014 [3]. The plan provided updated technical, financial and operational data for all WWTWs in the Gauteng Province, including the three plants within Municipality, namely, the Sebokeng, Rietspruit and Leeukuil WWTWs. At the time, the Municipality indicated that all WWTWs had proper maintenance programmes in place and rated their plants as "regular". However, they did indicate that the main challenge with the treatment plants was that they could not replace the components which were due for such, thus causing operational difficulties. Further, the Municipality had not provided feedback for the number of breakdowns they were experiencing at the time.

The report further confirmed the findings of 2008 that the WWTWs in the municipality had a total design capacity of approximately 172 ML/day and the incoming flow of some 208 ML/day. This demonstrated that the treatment capacity of the plants within the Municipality was inadequate, with two of the plants operating beyond their capacity, specifically, the Sebokeng WWTW was operating at 135 ML/day instead of 100 ML/day, and the Leeukuil WWTW at 45 ML/day instead of 36 ML/day. The Municipality indicated that the construction of the works to increase the capacity of the plants by upgrades or extensions had been delayed by some 4 to 6 years and that by the time those upgrades were completed new upgrades would be due [3].

3.3 The public outcry and the Human Rights Commission in 2018 to 2021

In 2018, following a public outcry, the Commission established an inquiry into the state of affairs with regard to water and sanitation management issues as well as the level and extent of pollution in the Vaal River System. The Commission, among other activities, undertook a visual inspection of the areas within the Municipality where complaints had emanated and publicised in different media outlets and platforms.

During the inspections the Commission discovered raw sewerage in a stream flowing through the Emfuleni Golf Course, burst sewer pipes on the banks of the Rietspruit River, dysfunctional components in the Rietspruit WWTW, blocked manholes, and children from a nearby school swimming in and consuming polluted water. Raw sewage was found to be discharging from dysfunctional wastewater treatment plants into the receiving tributaries of, and into the Vaal River [2].

The final report of the Commission was released in February 2021. In the findings the Commission found that the Vaal River was polluted beyond acceptable levels, which had serious implications for the Municipality as a water services authority [2]. Water services authorities have an obligation to protect water resources and to ensure that that residents have access to clean drinking water.

4 THEORETICAL CONSIDERATIONS

4.1 Basic design aspects of a wastewater collection and conveyance network

The basic design criteria for a wastewater collection network entails vertical and horizontal alignment design. The vertical alignment design requires that minimum grade requirements be met for self-cleansing capacity of the sewage being conveyed through the pipeline network. The flow through the network is ideally driven by gravity towards the wastewater treatment facility. However, should there be low points along the gravity main, it becomes necessary to install pump stations where sewage is pumped up a rising main to a high point where it can flow by gravity again until it reaches the treatment facility. Horizontal alignment design entails ensuring that change of direction at points of intersection are appropriate to ensure the natural flow of sewage. At points of change in grade and/or direction manholes need to be installed. It is critical that the pipe lengths in between the manholes be dead straight to avoid any potential blockages. The manholes are important to ensure that any blockages in the sewer network can be cleared when they occur.

The collapse of the wastewater collection network in the case of the Municipality points to lack of proper operation and maintenance by way of clearing the blockages when they occur in the collection network and failure to keep the pump stations in an operable condition by way of routine maintenance and replacement of mechanical and electrical components when required. This situation is further demonstrated in Section 5 below which deals with the status of the pollution problem from 2018 to date.

4.2 Asset management principles

The problem of water pollution in rivers, such as is the case in this study, emanates from the lack of the application of basic principles of integrated asset management, which is defined as, "an integrated process of decision-making, planning and control over the acquisition, use, safeguarding and disposal of assets to maximize their service delivery potential and benefits, and to minimise their related risks and costs over their entire life" [6]. This entails the appropriate development and implementation of such strategies and plans to operate, maintain, repair, refurbish or renew infrastructure assets, as well as making provision for the replacement of such assets or parts thereof when they have reached the end of their useful life.

Failure of the WWTWs to treat the wastewater to meet the legislated standards for discharge into the receiving environment is brought about by failure to apply the appropriate operation and maintenance standards and asset management principles.

In the case under study, indications are that appropriate investigations and condition assessments of the sanitation infrastructure assets were conducted and master planning was also carried out by the relevant authorities [3], [5]. What seems to have happened is a typical South African problem of developing impressive strategies and plans, but lack the appropriate implementation thereof.

5 FINDINGS: STATUS OF THE VAAL RIVER POLLUTION

According to a submission by the Department of Environmental Affairs to the Commission, water pollution in the country is prevalent with such problems having been identified in nine other water facilities outside the Vaal River system [2].

Although this case study focuses on the problems in the Emfuleni LM, pollution problems in the Vaal River System extend beyond the study area. For example, it was reported that a tributary to the Vaal River in Standerton, about 150 km upstream of the study area, also received untreated sewage from a dysfunctional wastewater management system [7]. It was also reported that a sewage spillage in Deneysville, about 40 km upstream from the study area, adjacent to the Vaal Dam wall [8].

The subsections below provide a snapshot of what was found in 2018 when the Commission initiated an Inquiry to the pollution problem in the Vaal River System, as well as some reported incidents in subsequent years.

5.1 Wastewater collection networks

Failure to maintain the sewage collection network by conducting scheduled or periodic maintenance and predictive or condition-based maintenance is evidently what led to the spillage of sewerage into the streets in residential areas as shown in Fig. 3. These spillages are not at low points in the sewer collection network, but at manholes. These blockages typically occur at points where the sewer main changes direction or grade or where there is an intersection of another sewer line. Such blockages would generally be reported to the municipality's water and sanitation operation and maintenance department to be cleared out. If there are no appropriate measures in place and the situation is not rectified timeously, one ends up with the situation that unfolded in the Municipality.

Figure 3: Sewage spillages in Sharpeville [9].

5.2 Pump stations

Failure to operate and maintain the pump stations according to sound asset management principles and best practice will invariably lead to sewage spillages at the pump stations. With the pump stations located at low points in the sewer networks, the sewage accumulates and ponds up without being pumped up and away from the pump stations. Fig. 4 shows typical cases where pump stations have failed as a result of which sewage created a pond inside and in the vicinity of the pump station. There are 44 pump stations in the Emfuleni sewer drainage basins or catchments, most of which have become dysfunctional.

Figure 4: A vandalised and abandoned pump station in Sharpeville and a flooded one in Vereeniging [9].

5.3 Wastewater treatment works

When the investigations into the nature and extent of the pollution problem in the Vaal were conducted, three wastewater treatment works were identified for prioritisation as part of the short-term intervention. In terms of the report released in 2014 [3], the three treatment plants had the status indicated as follows: The Sebokeng WWTW with a design capacity of 100 ML/day had inflows of 135 ML/day), Rietspruit WWTW with a design capacity of 36 ML/day had inflows of 28 ML/day) and the Leeukuil WWTW with a design capacity of 36 ML/day had inflows of 45 ML/day). All three treatment plants were operated on a 24 hour per day basis [3]. With the exception of the Rietspruit WWTW, the three plants were operated beyond their design capacity. However, the report indicated that all the plants met the downstream receiving body objectives at 100%. In terms of the downstream receiving body ecological status, only the Leeukuil WWTW was rated as good, and the Sebokeng and Rietspruit WWTWs were rated as fair [3].

The performance of the three wastewater treatment plants seems to have deteriorated to the extent that by 2018 there was a pollution crisis in terms of the water discharges coming out of the treatment plants. The pollution problem seems to be continuing even as the short-term intervention is being implemented because reports from as recently as October 2021 show that there is little improvement in the situation as shown in Fig. 5 [10]. The treatment works would be overloaded as the wastewater collection network blockages get cleared.

Figure 5: Images taken on 25 October 2021 showing effluent discharged into a canal from the Rietspruit WWTW and the pollution of the Rietspruit River 2 km downstream from the WWTW [10].

6 FINDINGS: INTERVENTIONS AND PROGRESS AS OF JANUARY 2022

There are two notable interventions that followed the process of the inquiry initiated by the Commission in 2018, namely, the deployment of the South African Defence Force and the implementation of the Sedibeng Regional Sanitation Scheme: Short Term Intervention, as further detailed below.

6.1 The South African Defence Force Intervention

Following the public outcry and the intervention by the Commission and other state oversight bodies, the South African Defence Force (SANDF) was duly deployed to the Municipality for the period from October 2018 to October 2019, which was further extended to January 2020 [11].

According to the presentation made to the Commission, the SANDF was deployed to provide assistance with engineering and other expertise to resolve the pollution crisis on the Vaal. However, because the SANDF engineers were not specialists in the repair and maintenance of wastewater treatment infrastructure, they had to contract external specialists to assist, for a period of seven days. The scope of work for the SANDF included prioritising the Sebokeng WWTW: three modules and four pump stations and sewer lines; the Rietspruit WWTW: two modules, two pump stations and sewage lines; and the Leeuwkuil WWTW: two modules, 36 pump stations and sewer lines [2]. The SANDF managed to clean most of the modules in the WWTWs and kept the pump stations regularly cleaned. This was, however, an immediate short-term intervention and a more permanent solution had to be developed and implemented. The SANDF highlighted challenges experienced during their intervention as the additional costs incurred during the operation and the lack of human resources within the Municipality which limited skills transfer [2].

6.2 The Sedibeng Regional Sanitation Scheme: Short-Term Intervention

Following the 2008 study into the capacity of the WWTWs in the Sedibeng DM, a regional scheme called the Sedibeng Regional Sanitation Scheme (SRSS) had been developed and proposed in a feasibility study released in 2010. According to a report released by the Department of Water Affairs in 2014, the SRSS was a Presidential Project as it was deemed to be critical for imminent growth in the region and to prevent the pollution of the Vaal River System through sewage spillages, a necessary intervention to protect the quality of drinking water sourced from the river system.

The report highly recommended that the matter highlighting the urgent need to implement the SRSS needed to be addressed at a top level between the Department of Water Affairs, Sedibeng DM, Midvaal LM and Emfuleni LM [3]. However, the report indicated that the Municipality and the neighbouring Midvaal LM had indicated that the projects were not being implemented at the time [3]. Without this recommendation being seriously considered and no action being taken, it was a matter of time before an ecological or environmental disaster happened.

When the situation finally got out of hand in 2018 and the sewage was seen flowing in the streets the authorities were forced to take drastic action. Among actions that were taken, as previously stated, was the deployment of the SANDF as a temporary solution. A more comprehensive intervention, albeit still short-term, was required for sustainable impact. This required the appointment of specialist engineers in wastewater treatment designs and implementation, particularly on the side of operation and maintenance of the wastewater management systems. Specialist and experienced civil, mechanical and electrical contractors were also required.

With the release of the final report of the Commission in February 2021, engineers were appointed for consultancy services for the short-term intervention, referred to as the Vaal River System Intervention programme (VRSI), as well as for the development and implementation of a medium to long term solution. This would include new infrastructure development and the optimisation and refurbishment of existing infrastructure.

The primary focus of the VRSI which was initiated in 2021 is to revitalise the existing wastewater collection, conveyance and treatment infrastructure. This is aimed at restoring the infrastructure to operational efficiency in the region. This intervention is part of the original SRSS envisaged in 2010 in terms of the scope of work for the medium to long term, but the current focus is on the short-term intervention which entails clearing up the wastewater collection network, clearing out the pump stations and treatment plants by carrying out remedial civil works and refurbishing and/or replacing the mechanical and electrical components [12].

The appointed civil engineers indicated that besides the refurbishment of the three WWTWs of Sebokeng, Rietspruit and Leeuwkuil, 44 individual pump stations are being upgraded, and the associated conveyance and sewer network pipelines are being unblocked. The DWS procured the services of Plant Hire and Bio-solids to unblock the sewer network and conveyance system and to handle bio-solids, respectively. The DWS also procured Civil Framework Contractors and aim to further procure the services of the Mechanical and Electrical (M&E) Contractors [12].

Progress on the implementation of the VRSI, as reported by the consulting engineers and the DWS, as of the end of January 2022 is outlined below.

6.2.1 Target for the quarter

The target for the last quarter of 2021 was to repair, refurbish and replace the various failed infrastructures, including bulk sewer lines, pumps, motors, electrical panels, screens, and manholes.

6.2.2 Actual progress for the quarter

In the last quarter of 2021, the actual progress included the completion of 195 high-pressure unblockings. Cleaning operations were logged daily in the Sebokeng Catchment, 101 in the Leeuwkuil Catchment and 100 in the Rietspruit Catchment. Cleaning operations in the three catchments are ongoing, which involves, among others, the cleaning of bulk lines over longer pipeline sections.

6.2.3 Reasons for non-achievement of targets

Some targets were not achieved because of the delays in concluding Service Level Agreements with the appointed seven (7) Civil Engineering Framework Contractors. The non-responsiveness of Mechanical Engineering framework contractors and the unavailability of Electrical Engineering Framework of Contractors also impacted performance [12].

Persistent community unrests or protests during implementation result in delays and increased costs of implementation. For example, the current intervention programme was disrupted when the bio-solids handling activities were halted from 13 October to 9 December 2021 in the Sebokeng WWTW and from 13 October to 13 December 2021 in the Rietspruit WWTW [12].

The community protest are due to the communities demanding job opportunities and the engagement of local enterprises beyond what is practicably possible on the project. At the end of March 2022, it was not possible to conduct a site visit to review progress because situation was said to be volatile.

7 CONCLUSION AND RECOMMENDATIONS

Some of the critical challenges that have led to environmental pollution of land and water resources in the Sedibeng DM region include the ageing sanitation infrastructure resulting in performance failures and a rise in maintenance costs.

Non-implementation of operation and maintenance strategies and plans led to a total collapse of the sanitation management system, resulting in the pollution levels being above acceptable standards, as the Commission found in their final report. Had the SRSS as first proposed in 2010 been implemented timeously, the crisis situation would not have arisen. As the crisis is continuing, it is important that all necessary resources, especially human capital and project funding, be made available. It would probably be helpful to reinstate the Presidential status of SRSS as was the case when it was first proposed in 2010.

The procurement processes for service providers and contractors needs to be prioritised and any bottlenecks cleared up in order to ensure timeous appointments and implementation to ensure successful completion of the intervention programme.

The problem of persistent community protests and unrests is a serious one and is responsible for many delays in the implementation of projects in general, and in particular, the implementation of the SRSS and the VRSI. A robust community engagement strategy and plan needs to be developed and implemented with the support of all levels of government and politicians to manage community expectations. Political promises which create unrealistic expectation from the community need to be avoided.

REFERENCES

[1] Sedibeng District Municipality, Annual Report 2010–2011, 2011. http://www.sedibeng.gov.za/a_keydocs/ar_2010_11/ar_2010_11_complete.pdf. Accessed on: 2 May 2022.

[2] The South African Human Rights Commission, Final Report of the Gauteng Provincial Inquiry Into the Sewage Problem of the Vaal River, pp. 38, 43–46, 2018. https://www.sahrc.org.za/home/21/files/The%20Vaal%20Inquiry%20Final%20Report_17022021.pdf. Accessed on: 25 Feb. 2022.

[3] Republic of South Africa, Department of Water Affairs, *Masterplan for Gauteng Wastewater Treatment Works. Book D – Sedibeng District Municipality*, Infraburo (Pty) Ltd and Inviromap CC, Government Printers: Pretoria, 2014.

[4] Republic of South Africa, Act No. 108 of 1997: Water Services Act, 1997, Government Gazette No. 18522, Section 1, Cape Town, 1997.

[5] Arcus GIBB Engineering & Science, Sedibeng Regional Sanitation Scheme: A study of current and expected future flowrates to determine the wastewater treatment capacity requirements of the region up to 2025, 2008. https://www.biblebeliever.co.za/Midvaal2/EIA_for_Midvaal_Area/Sedibeng%20Regional%20Sanitation%20Scheme/J1184%20Gaut%20002_09-10_N0094%20SRSS%20CURRENT%20%20FUTURE%20FLOWS.pdf. Accessed on: 25 Feb. 2022.

[6] Mnguni, E.S., Water infrastructure asset management: A comparative analysis of three urban water utilities in south Africa. *WIT Transactions on Ecology and the Environment*, vol. 217, WIT Press: Southampton and Boston, pp. 927–938, 2019.

[7] Schoombee, M., Standerton's unrelenting pollution of the Vaal River has dire consequences. *Ridge Times*, 8 Dec. 2020. https://ridgetimes.co.za/248887/standertons-unrelenting-pollution-of-the-vaal-river-has-dire-consequences/. Accessed on: 20 Feb. 2022.

[8] *Vaalweekblad*, Polluting Gauteng's drinking water, 16 Jul. 2015. https://vaalweekblad.com/2856/polluting-gautengs-drinking-water/ Accessed on: 20 Feb. 2022.

[9] Organisation Undoing Tax Abuse, Emfuleni washes away Human Rights: A submission by the OUTA to the Human Rights Commission, 25 Sep. 2018. https://www.outa.co.za/web/content/76311. Accessed on: 30 Apr. 2022.

[10] Evans, J., 2021. "A story of South Africa": Emfuleni residents fed up with Vaal River pollution inertia. *Daily Maverick*, 29 Oct. 2021. https://www.dailymaverick.co.za/article/2021-10-29-a-story-of-south-africa-emfuleni-residents-fed-up-with-vaal-river-pollution-inertia/#:%7E:text=Lik%E2%80%A6. Accessed on: 25 Jan. 2022.

[11] South African Government, Media statement: Defence withdraws from Vaal River Intervention Project, 2020. https://www.gov.za/speeches/vaal-river-intervention-project%C2%A0-27-jan-2020-0000. Accessed on: 2 May 2022.

[12] GIBB Infrastructure and Transport, Sedibeng Regional Sanitation Scheme: Monthly progress report for the Department of Water and Sanitation (DWS) for the Vaal River system intervention (refurbishment works), Period ending 31 Jan. (unpublished), 2022.

ENVIRONMENTAL IMPACTS OF REVERSE OSMOSIS IN WASTEWATER TREATMENT VERSUS DESALINATION TO MEND THE WATER CYCLE: A LIFE CYCLE ASSESSMENT

ESRA ALEISA[1], ANWAR ALJUWAISSERI[2], KHAWLA ALSHAYJI[3] & ASMAA AL-MUTIRI[4]
[1]Industrial and Management Systems Engineering Department, Kuwait University, Kuwait
[2]Kuwait Petroleum Corporation, Kuwait
[3]Chemical Engineering Department, Kuwait University, Kuwait
[4]Department of General Services, Ministry of Electricity and Water, Kuwait

ABSTRACT

Sustainable water production is one of the top five challenges facing humanity within the upcoming decade; for arid regions, this challenge is aggravated. In this research, we evaluate the fastest growing technology in water treatment, reverse osmosis (RO), for both municipal wastewater treatment and seawater desalination to combat the challenges of water scarcity and climate change. We conduct a life cycle assessment (LCA) to evaluate the environmental impacts of municipal wastewater treatment and seawater desalination for a consistent functional unit using ISO 14040/44 standards. The modeling concept adopts a cradle-to-gate consequential paradigm. The life cycle inventory is based on field data collection from one of the largest wastewater RO plants worldwide, as well as reports, literature and ecoinvent database processes. The life cycle impact assessment is conducted on both the characterized and normalized levels using the ReCiPe method. The results are intended to assist policy-makers in better managing water resources. The study is applied to Kuwait but has wider repercussions.

Keywords: reverse osmosis, desalination, wastewater, life cycle assessment (LCA), Gulf Cooperation Council (GCC), Kuwait.

1 INTRODUCTION

The UN water report [1] has shown that the Gulf Cooperation Council (GCC) states are the top countries worldwide in terms of the water scarcity index (Fig. 1). The increasing population, changing lifestyles and warmer weather due to climate change have resulted in a sixfold growth in global water demand over the past century [1]. For Kuwait, although the country's commitment to the Sustainable Development Goal (SDG) No. 6 of clean water and sanitation is commendable, the price for this achievement has affected other SDGs. To name one, SDG No. 13 of climate action, as potable water is mainly produced through thermal desalination, an energy-intensive process. In the GCC, a staggering 30%–50% of oil production is consumed by the cogeneration of electricity and desalination. The COP21 in Glasgow, UK and the figures of The 2019 UN Environment Programme's Emissions Gap Report indicate that not enough action is taken worldwide to mitigate carbon emissions [2]. The report concludes that to abide by the Paris Agreement, emissions need to drop by 7.6% annually by 2030 for the 1.5°C target. For this seemingly "impossible" task, the European Union (EU) launched a new world-challenging mission to reach the objectives of the Paris Agreement to reduce global warming to preindustrial era levels, the "Mission possible". For the same urgent cause, the United Nations has also launched the "Decade of Change" mission to mobilize everyone everywhere to increase momentum and take immediate action and gather forces to deliver the 2030 promise [3]. It has long been agreed that energy efficiency at the end-use stage is the fastest and most cost-effective solution for reducing greenhouse emissions [4]. Countries worldwide have established targets to achieve sustainable water and energy systems, consistent with the SDGs [5].

WIT Transactions on Ecology and the Environment, Vol 257,
www.witpress.com, ISSN 1743-3541 (on-line)
doi:10.2495/WMEI220031

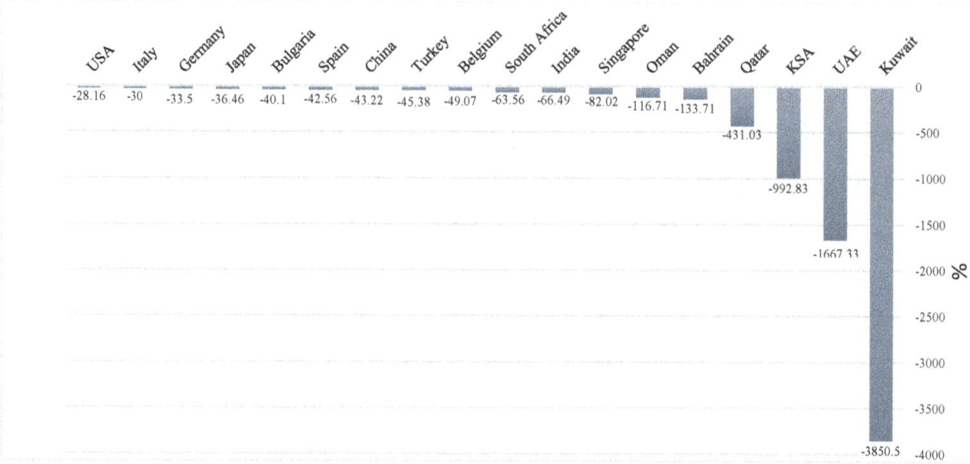

Figure 1: Water stress (%) for GCC countries: Bahrain, Kuwait, Oman, Qatar, the Kingdom of Saudi Arabia (KSA) and the United Arab Emirates (UAE) in comparison with selected countries [6].

Our area of focus is the most arid area worldwide, the GCC [7]. The GCC presents an intricate paradox of severe water scarcity coupled with staggered water consumption, which is a result of rentierism and socioeconomic situations. The GCC regions have long relied on costly thermal seawater desalination, followed by the high abstraction of nonrenewable groundwater resources, to satisfy their demand for water [8]–[10]. The availability of fossil fuels at low extraction costs has slowed the conversion to energy-efficient desalination in the GCC, such as multistage flash distillation (MSF) [10], [11]. Currently, desalination megaprojects in the GCC are transitioning to membrane-based desalination plants (DPs) [12]–[14]. For instance, the Yanbu seawater reverse osmosis (SWRO) DP in the Kingdom of Saudi Arabia (KSA) is one of the largest SWRO plants in the Middle East. The plant started its operation in 2020 with a total capacity of 450K m^3/day [15]. Similarly, the Taweelah SWRO DP is expected to start its operation in 2022 with a capacity of 450,000 m^3/day [15].

Kuwait has three SWRO DPs: the Shuwaikh SWRO DP and the Al-Zour South SWRO DP, which commenced in 2013 and 2014, respectively, and the Doha SWRO DP, which was recently inaugurated but was not included and has not yet reached its design capacity. The total installed capacity of each of the aforementioned operating SWRO DPs is 30 million imperial gallons per day (MIGPD) [11]. This number is expected to increase when the Doha SWRO DP reaches its full operation capacity. The Shuwaikh and Al-Zour South SWRO DPs combined contribute only 12% of the total freshwater production in the country (see Fig. 2), while the rest is produced using MSF and the newly inaugurated multieffect distillation (MED) DP [9].

Wastewater reuse is now considered indispensable for meeting the increasing water demand, particularly under conditions of alarming water scarcity, which are now already affecting every continent [16]. The GCC has already embarked on constructing mega wastewater treatment plants (WWTPs) and projects. The overall capacity of WWTPs in the GCC is approximately 2,800 Mm^3/yr (million cubic meters per year). The total number of WWTPs is 241; these stations primarily use secondary and tertiary methods [8]. One

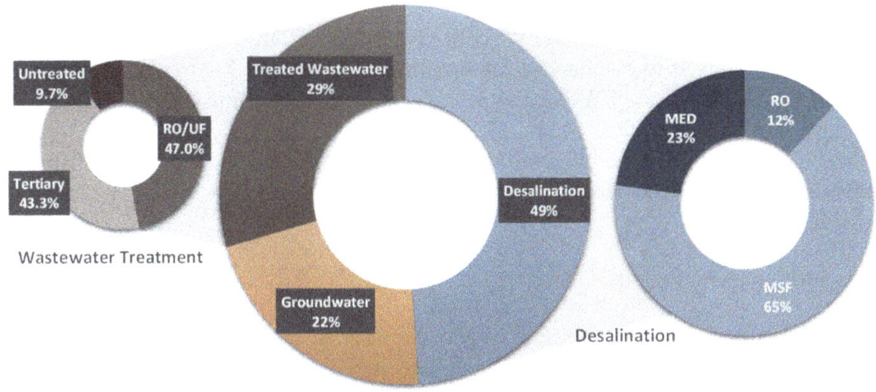

Figure 2: Water sources in Kuwait.

exception to this is in Kuwait, which vastly uses RO to treat wastewater (WWRO). Over 90% of Kuwait municipal wastewater is treated, of which 47% is treated through ultrafiltration (UF) and RO, and the remaining is treated through tertiary and advanced tertiary treatment (see Fig. 2). There are seven WWTPs in Kuwait: Alriqqa, Um Alhayman, Sulaibiya, Kabd, Wafra, Sabah Alhamed and Alkhiran (pilot plant). All WWTPs treat wastewater to the tertiary level, except for the Sulaibiya WWTP, which uses UF/RO to achieve potable water quality for nonpotable use [17].

This study conducts a life cycle assessment (LCA) on state-of-the-art RO treatment used for both seawater desalination and domestic wastewater treatment. The results are intended to assist policy-makers in better utilizing conventional water resources in Kuwait. The study is applied to Kuwait but has wider repercussions.

2 ASSESSING RO FOR WATER DESALINATION AND WASTEWATER TREATMENT USING LCA

LCA has been implemented to assess the environmental impacts of DPs with different plant characteristics [9], [18]. Al-Shayji and Aleisa [11] investigated the environmental impacts of all operating DPs in Kuwait. Mannan et al. [19] highlighted the regional impact on the environmental impact of a DP. Aljuwaisseri et al. [9] investigated the impacts of intake water salinity and turbidity on the environmental impact. Researchers have addressed varying desalination process scenarios, including different gain ratios [19], multiple energy sources [10], [20], [21] and different capacities [22]. The effects of using renewable energy sources in DPs and their associated impacts on environmental loads have also been addressed within LCAs [19], [20], [23]. A literature review on SWRO LCAs can be found in Aljuwaisseri et al. [9].

For wastewater treatment assessment using LCAs, to date, more than 100 research papers have been published in this field [24]. Detailed literature reviews comparing different LCA objectives, challenges, methodological choices and results related to wastewater treatment and sludge management can be found in Corominas et al. [24], Yoshida et al. [25], Pradel et al. [26], and Gallego-Schmid and Tarpani [27]. In addition, LCA has been used to analyze environmental impacts in the field of storm-water management [28], to determine appropriate solutions in the field of the urban water cycle [29], to control emitted greenhouse gases [30], and to identify the environmental impacts of wastewater sludge treatment [31].

3 MATERIALS AND METHODS

The goal of this study is to evaluate the environmental burden of RO treatment on two main water production processes, SWRO and WWRO, using LCA. Both analyses adhere to the four stages outlined by ISO 14044 [32]. The analysis is conducted using open-loop consequential modeling. The UF used is 1 Mm^3 of permeate to potable quality per the requirements of the World Health Organization (WHO). The two system boundaries are described next.

3.1 System scope and boundary

The system boundaries are the cradle-to-gate impacts of all processes, materials, energy requirements and chemicals through operation and are calculated through field visits, reports and literary analysis. Water delivery and disposal are excluded. The construction and decommissioning phases were excluded from the LCA due to their insignificant contributions to the total environmental impact [20], [33]–[35]. For instance, Raluy et al. [36] found that the construction phase only contributed 5% to the total environmental impact. Furthermore, brine disposal was also excluded from the study due to its relatively negligible environmental impact [34], [36].

3.1.1 SWRO system description

The process of producing desalinated water using SWRO is based on Aljuwaisseri et al. [9] and is shown in Fig. 3. Ferric chloride is injected to coagulate the particles, improving the removal of suspended material through gravity filters. Sodium bisulfate is injected to remove the residual chlorine to protect the membranes from oxidization. An antiscalant is also

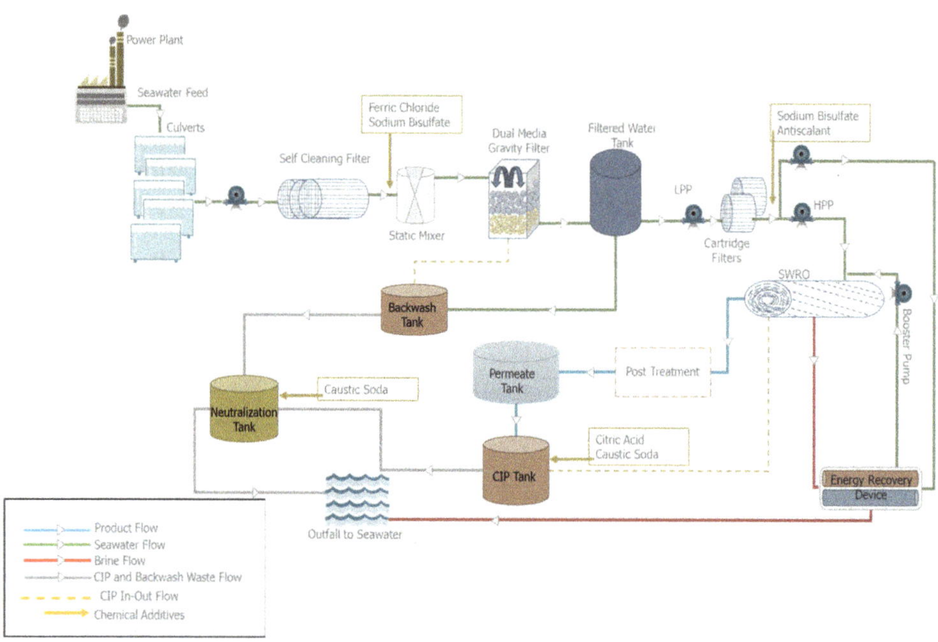

Figure 3: Process flowchart for the SWRO DP. *(Source: Adapted from [9].)*

injected into the filtered seawater before it enters the SWRO unit. The SWRO system is a single-pass system with energy-recovery device units supplied with filtered seawater by a low-pressure pump. The concentrated brine rejection line is located at the outlet of the membrane array to feed the energy-recovery device. Finally, the permeate is mixed with the distilled water produced at a nearby plant before it is sent to the distribution network. Clean-in-place (CIP) is the main cleaning process applied to remove scaling and biofouling. Citric acid and caustic soda are the chemicals used in CIP.

3.1.2 UF/RO system description

Wastewater first undergoes a primary treatment that is a physical/mechanical process that removes suspended and floating particles from the wastewater entering the WWTP. Primary treatment includes screening to screen grit and other suspended solids. Secondary treatment uses biological processes to digest and dissolve organic pollutants to produce settleable solids. The process is followed by aeration basins or settling tanks to clarify the influent by removing approximately 85% of its suspended solids and BOD [37], [38]. The biological treatment stage uses a vertical loop reactor (VLR) [39]. The wastewater is then moved by gravity to secondary wastewater treatment equipment, which consists of aeration chambers and primary clarifiers. The type of clarifier used is a rim flow clarifier. The resultant active sludge is continuously recycled to the aeration tank. The surplus activated sludge (SAS) is used for sludge treatment [31]. The effluent undergoes a tertiary treatment, which eliminates over 95% of all impurities from the sewage. It consists of rotating disc filters and an ultraviolet (UV) system. Advanced treatment uses UF/RO as the membrane process to remove the residual solids, inorganics, organics and microorganisms remaining in the tertiary effluent. The UF/RO treatment is built based on the Sualibiayh WWTP. The UF system consists of 8.7 K UF membranes, and the RO trains consist of 21 K membranes. Eventually, 85% of the tertiary effluent is reclaimed, and the remaining 15% is discharged as brine into the sea [40].

Excess sludge is thickened to reduce its volume. A polymer preparation unit (PPU) is used for additional thickening and flocculation [39]. Aerobic digesters use gravity belts that carry sludge for dewatering to form a sludge cake, which is then landfilled [31].

The chemical and microbiological characteristics of the wastewater RO permeate (ROP) in Kuwait exceed the WHO standards [17], [41], [42] of potable quality and the Kuwait standards as published by the Kuwait Environmental Public Authority (KEPA) and the Ministry of Public Works (MPW) (see Table 1).

Table 1: Specifications of the effluent per cubic meter [43].

Specification	**Unit**	**ROP**	**Potable water**	**Specification**	**Unit**	**ROP**
pH	–	6–8	6.8–7.5	Chloride	mg/L	
Conductivity			515	Ammonia	mg/L	< 1
TSS	mg/L	< 1	–	Nitrite	mg/L	< 1
VSS	mg/L	< 1	–	Total count	Colony/100 mL	Nil
COD	mg/L		–	T. Coli	Colony/100 mL	Nil
BOD	mg/L	< 1	–	F. Coli	Colony/100 mL	Nil
Grease and oil	mg/L	< 0.05	–	Salmonella	Colony/100 mL	Nil
TDS	mg/L	< 100	400	Streptococci	Colony/100 mL	Nil

Note: BOD = biochemical oxygen demand; COD = chemical oxygen demand; TSS = total suspended solids; VSS = volatile suspended solids; T. Coli = total coliforms; F. Coli = fecal coliforms.

3.2 Life cycle inventories

The life cycle inventory (LCI) foreground data, including energy consumption, chemical dosage, and membrane data, were collected from field visits and reports. The background data comprising the production of chemicals, energy and membranes were obtained from [44] and the ecoinvent database v.3.0. For SWRO, the LCI is provided in Table 2. The tertiary wastewater treatment LCI is provided in Table 3. The UF/RO LCI was obtained from the Sulaibiya WWTP in Kuwait [40]. The energy calculations were obtained from Aleisa and Heijungs [10] and Aleisa and Al-Shayji [21]. The electrical energy required for SWRO is 5.2 kWh/m^3 [9]. The electrical energy required for tertiary wastewater treatment is 3.95×10^{-1} kWh/m^3, and that required for UF/RO (excluding tertiary) is 3.14×10^{-1} kWh/m^3 [40].

Table 2: LCI of the SWRO DP. *(Source: Adapted from [9], [44].)*

Input data	Formula	Unit	Amount	Ecoinvent process
Antiscalant hydrex	$CH_2 = CHCOOH$	g/m^3	15.857	Acrylic acid [9]
Caustic soda	$NaOH$	g/m^3	56.2	Sodium hydroxide 50% in solution [44]
Ferric chloride	$FeCL_3$	g/m^3	45.31	Iron (III) chloride 40% in solution [44]
Sodium bisulfate	$NaHSO_3$	g/m^3	0.906	Sodium hydrogen sulfite [9]
Citric acid	$C_6H_8O_7$	g/m^3	0.906	Citric acid [9]
SWRO membrane		g/m^3	0.898	Glass fiber-reinforced plastic, polyamide [9]

Table 3: Chemical additives per cubic meter of tertiary treated wastewater [40].

Chemicals	Formula	Amount (g)
Sodium hydroxide 50%	$NaOH$	1.096
Sodium hypochlorite 12.5%	$NaOCl$	2.740
Activated carbon	C	3.044×10^{-2}
Cationic polymer	–	1.461
Chlorine liquid	Cl	3.288

The landfilling facility is designed for biogenic waste from the ecoinvent database version 3.0. It has a design capacity of 1.8 million m^3 with a 30-year lifetime. It is equipped with a leachate and landfill gas collection system [31], [45].

3.3 Life cycle impact assessment

The life cycle impact assessment (LCIA) was calculated using the ReCiPe midpoint (H) V1.10. Table 4 lists the midpoint impact categories.

4 RESULTS

Figs 4 and 5 illustrate the characterized and normalized LCIA results, respectively. The LCIA shows a significant difference in the environmental impacts between SWRO and WWRO. Using the characterized results (see Fig. 4), SWRO scores higher (worse) in terms of CC, OD, PMF and FD. This was mainly due to the high energy consumption of SWRO. Over

Table 4: Midpoint impact categories investigated in the LCIA phase (ReCiPe V1.10).

Midpoint impact category	Unit
Climate change (CC)	kg (CO_2 to air)
Ozone depletion (OD)	kg (CFC-11 to air)
Fossil depletion (FD)	kg (oil)
Metal depletion (MD)	kg (Fe)
Particulate matter formation (PMF)	kg (PM_{10} to air)
Human toxicity (HT)	kg (14-DCB to urban air)
Marine ecotoxicity (MET)	kg (14-DCB to marine water)

92% of the environmental burden is attributed to the electricity use required by SWRO. Ferric chloride followed by the antiscalant contributed 3% and 1%, respectively, to the environmental impact of the SWRO DP. The process contribution analysis revealed that petroleum and gas production were the highest contributors to FD (95% contribution). The main contributor to CC was the CO_2 resulting from high-voltage electricity production. On the other hand, WWRO scores higher (worse) in characterized values in terms of MET and MD. This is due to the extensive use of ferric chloride and hydrochloric acid in WWRO. The results are similar in terms of HT at 52.7% and 53.3% for WWRO and SWRO, respectively. The first is due to ferric chloride, and the second is due to energy. The normalized results (see Fig. 5) indicate that FD is the impact category with the largest effect, followed by CC, MD and then PMF.

The single score results (see Fig. 6) indicate that the WWRO impact is only 31% of that of the SWRO impact. Regulations exclude all amenity uses of WWRO and restrict its uses to the irrigation of crops and some industrial applications [17]. 19% of all water consumed in the agricultural sector is recycled water. WWRO treatment is regulated by KEPA standards [46], [47], which are on several parameters more conservative than those of the WHO [41]. No outbreaks of infectious disease have occurred since 2005, when the first utilization of

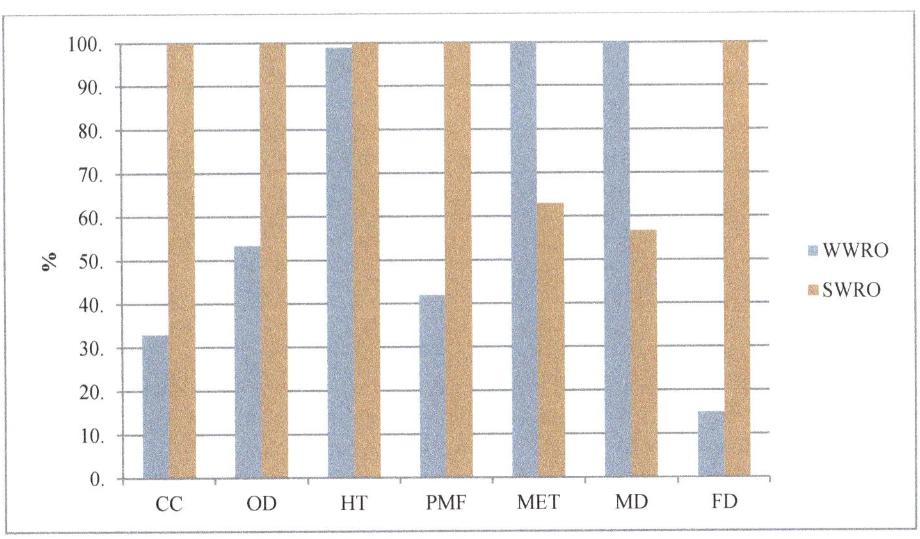

Figure 4: Characterized results for WWRO versus SWRO production using ReCiPe Midpoint (H) V1.10.

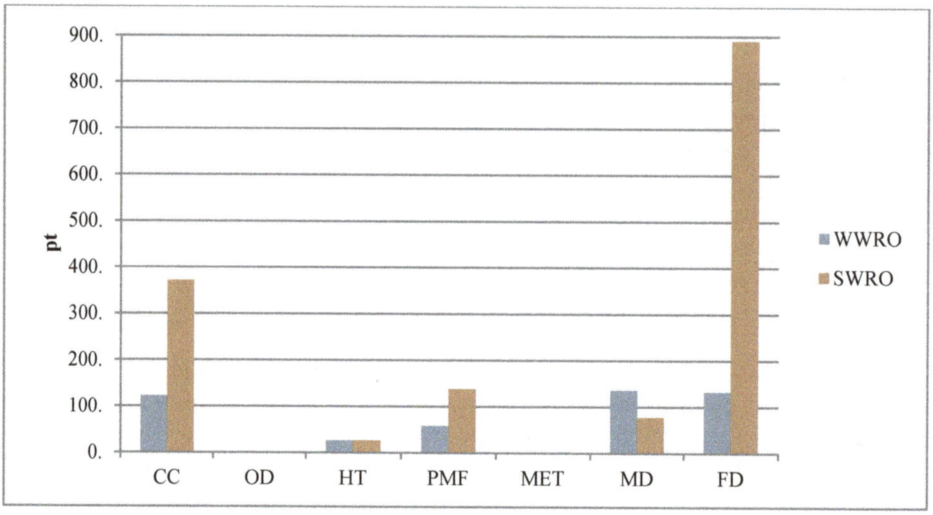

Figure 5: Normalized results for WWRO versus SWRO production using ReCiPe Midpoint (H) V1.10.

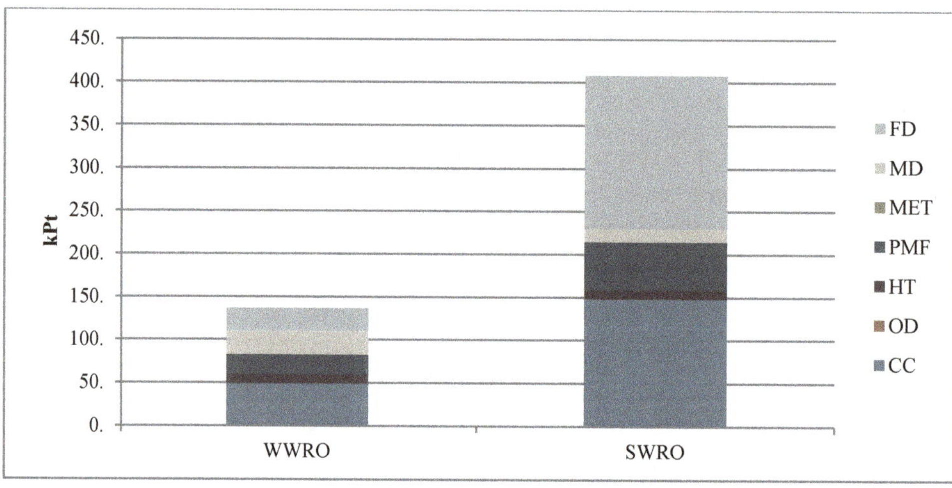

Figure 6: Single score results for WWRO versus SWRO production using ReCiPe Midpoint (H) V1.10.

ROP water took place. This clearly indicates a great opportunity that could significantly contribute to balancing SDG No. 6 (clean water and sanitation) and SDG No. 7 (affordable and clean energy) to contribute to SDG No. 13 (climate action).

5 CONCLUSIONS

This study provides an LCA that compares the environmental impacts of WWRO versus SWRO. SWRO has three times the environmental impact of WWRO. This result was most significant in terms of CC, OD, PMF and FD. This was mainly due to the high energy consumption of SWRO. Over 92% of the environmental burden is attributed to the electricity use required by SWRO.

On the other hand, WWRO scores higher (worse) in characterized values in terms of MET and MD. This is due to the extensive use of ferric chloride and hydrochloric acid in WWRO. The local regulations exclude all amenity uses of WWRO and restrict its uses to the irrigation of crops and some industrial applications. Relaxing some of these regulations could significantly and safely contribute to a circular economy and promote a better balance between achieving SDG No. 6 (clean water and sanitation) and SDG No. 13 (climate action).

REFERENCES

[1] UN Water, *United Nations World Water Development Report 2020: Water and Climate Change*, ed. R. Connor, UN Water: Paris, 2020.

[2] UNEP, Emissions Gap Report 2019. United Nations Environment Programme: Nairobi, 108 pp., 2019.

[3] UN.org, Decade of change. 2020. https://www.un.org/sustainabledevelopment/decade-of-action/. Accessed on 16 Sep. 2020.

[4] Rezessy, S., Bertoldi, P. & Voogt, M., Mission possible: Bringing end-use energy efficiency to the European Emissions Trading scheme. Summer study of the European Council for Energy Efficient Economy. ECEEE: La Colle Sur Loup, 2007.

[5] Nock, D. & Baker, E., Holistic multi-criteria decision analysis evaluation of sustainable electric generation portfolios: New England case study. *Applied Energy*, **242**, pp. 655–673, 2019.

[6] Mateo-Sagasta, J. et al., *Safe Use of Wastewater in Agriculture*, eds J. Liebe & R. Ardakanian, Food and Agriculture Organization (FAO) of the United Nations and United Nations Water: Bonn, 2016.

[7] GCC Secretariat General, Development of a Unified Water Sector Strategy for Gulf Cooperation Council of Arab Member States (DRAFT) (English). King Abdullah Institute for Research and Consulting Studies: Riyadh, 2015.

[8] Aleisa, E. & Al-Zubari, W., Wastewater reuse in the countries of the Gulf Cooperation Council (GCC): The lost opportunity. *Environmental Monitoring and Assessment*, **189**(11), pp. 553–568, 2017.

[9] Aljuwaisseri, A., Aleisa, E. & Alshayji, K., Environmental and economic analysis for desalinating seawater of high salinity using reverse osmosis: A life cycle assessment approach. *Environment, Development and Sustainability*, 2022.

[10] Aleisa, E. & Heijungs, R., Leveraging life cycle assessment and simplex lattice design in optimizing fossil fuel blends for sustainable desalination. *The International Journal of Life Cycle Assessment*, **25**(4), pp. 744–759, 2020.

[11] Al-Shayji, K. & Aleisa, E., Characterizing the fossil fuel impacts in water desalination plants in Kuwait: A life cycle assessment approach. *Energy*, **158**, pp. 681–692, 2018.

[12] IRENA, *Renewable Energy Market Analysis*, IRENA: Abu Dhabi, UAE, 154, pp., 2019.

[13] Kaya, A., Tok, M.E. & Koc, M., A levelized cost analysis for solar-energy-powered sea water desalination in the Emirate of Abu Dhabi. *Sustainability*, **11**(6), pp. 1691, 2019.

[14] Mena desalination market, Ventures ONSITE, 2019. menadesal.com.

[15] Project, B.f.M.D., Major Desalination Plants in MENA, p. 4, 2019.

[16] UN Water, Water scarcity, 2020. https://www.unwater.org/water-facts/scarcity/.

[17] Aleisa, E. & Alshayji, K., Analysis on reclamation and reuse of wastewater in Kuwait. *Journal of Engineering Research*, **7**(1), pp. 1–13, 2019.

[18] Aziz, N. & Hanafiah, M.M., Application of life cycle assessment for desalination: Progress, challenges and future directions. *Environ. Pollut.*, **268**(Pt B), 115948, 2021.
[19] Mannan, M. et al., Examining the life-cycle environmental impacts of desalination: A case study in the State of Qatar. *Desalination*, **452**, pp. 238–246, 2019.
[20] Shahabi, M.P. et al., Environmental life cycle assessment of seawater reverse osmosis desalination plant, powered by renewable energy. *Renewable Energy*, **67**, pp. 53–58, 2014.
[21] Aleisa, E. & Al-Shayji, K., Ecological–economic modeling to optimize a desalination policy: Case study of an arid rentier state. *Desalination*, **430**, pp. 64–73, 2018.
[22] Bhakar, V. et al., Life cycle assessment of filtration systems of reverse osmosis units: A case study of a university campus. *13th Global Conference on Sustainable Manufacturing: Decoupling Growth from Resource Use*, eds G. Seliger, H. Kohl & J. Mallon, pp. 268–273, 2016.
[23] Al-Kaabi, A.H. & Mackey, H.R., Environmental assessment of intake alternatives for seawater reverse osmosis in the Arabian Gulf. *Journal of Environmental Management*, **242**, pp. 22–30, 2019.
[24] Corominas, L. et al., Life cycle assessment applied to wastewater treatment: State of the art. *Water Research*, **47**(15), pp. 5480–5492, 2013.
[25] Yoshida, H., Christensen, T.H. & Scheutz, C., Life cycle assessment of sewage sludge management: A review. *Waste Manag. Res.*, **31**(11), pp. 1083–101, 2013.
[26] Pradel, M. et al., From waste to added value product: Towards a paradigm shift in life cycle assessment applied to wastewater sludge: A review. *Journal of Cleaner Production*, **131**, pp. 60–75, 2016.
[27] Gallego-Schmid, A. & Tarpani, R.R.Z., Life cycle assessment of wastewater treatment in developing countries: A review. *Water Research*, **153**, pp. 63–79, 2019.
[28] Tavakol-Davani, H. et al., Combining hydrologic analysis and life cycle assessment approaches to evaluate sustainability of water infrastructure. *Journal of Irrigation and Drainage Engineering*, **144**(11), 2018.
[29] Petit-Boix, A. et al., Addressing the life cycle of sewers in contrasting cities through an eco-efficiency approach. *Journal of Industrial Ecology*, **22**(5), pp. 1092–1104, 2018.
[30] Raghuvanshi, S. et al., Comparative study using life cycle approach for the biodiesel production from microalgae grown in wastewater and fresh water. *Procedia CIRP*, **69**(1), pp. 568–572, 2018.
[31] Aleisa, E., Alsulaili, A. & Almuzaini, Y., Recirculating treated sewage sludge for agricultural use: Life cycle assessment for a circular economy. *Waste Management*, **135**, pp. 79–89, 2021.
[32] ISO 14044, Environmental Management – Life Cycle Assessment – Requirements and Guidelines. The International Organization for Standardization (ISO): Geneva, 2006.
[33] Garfi, M. et al., Life cycle assessment of drinking water: Comparing conventional water treatment, reverse osmosis and mineral water in glass and plastic bottles. *Journal of Cleaner Production*, **137**, pp. 997–1003, 2016.
[34] Tarnacki, et al., Comparison of environmental impact and energy efficiency of desalination processes by LCA. *Water Science and Technology-Water Supply*, **11**(2), pp. 246–251, 2011.
[35] Zhou, J., Chang, V.W.C. & Fane, A.G., Life cycle assessment for desalination: A review on methodology feasibility and reliability. *Water Research*, **61**, pp. 210–223, 2014.

[36] Raluy, G., Serra, L. & Uche, J., Life cycle assessment of MSF, MED and RO desalination technologies. *Energy*, **31**(13), pp. 2361–2372, 2006.

[37] Hamoda, M., Advances in wastewater treatment technology for water reuse. *Journal of Engineering Research*, **1**(1), pp. 1–27, 2013.

[38] Hsu, A. et al., Environmental Performance Index: Full Report and Analysis. Yale Center for Environmental Law and Policy, Yale University: New Haven, CT, 2014.

[39] MPW, Kabd WWTP, S.E. sector, Department of Operation and Maintenance of Plants (North Zone): Kuwait, 2019.

[40] Aleisa, E., Al-Mutairi, A. & Hamoda, M.F., Reconciling water circularity through reverse osmosis for wastewater treatment for a hyper-arid climate: A life cycle assessment. *Sustainable Water Resources Management*, 2022.

[41] WHO, A compendium of standards for wastewater reuse in the Eastern Mediterranean Region C.f.E.H.A. (CEHA), 2006.

[42] FAO, Wastewater treatment and human exposure control: Kuwait. Wastewater use case studies, 1992.

[43] MPW, Reports from the sanitary engineering sector in Kuwait. Ministry of Public Work: Kuwait, 2018.

[44] Vince, F. et al., LCA tool for the environmental evaluation of potable water production. *Desalination*, **220**(1–3), pp. 37–56, 2008.

[45] Aleisa, E. & Heijungs, R., Leveraging life cycle assessment to better promote the circular economy: A first step using the concept of opportunity cost. *Sustainability*, **14**(6), 2022.

[46] KEPA, Kuwait Environmental Law 42/2014 emended by law 99/2015. K.E.P. Authority, Kuwait Al Youm: Kuwait, 2014.

[47] KEPA, Decree No. 12 of 2017 issuing the Executive Regulations for the Protection of the Aquatic and Coastal Environment from Pollution (Articles 88, 90, 92 and 94–99 of Law No. 42/2014), K.E.P. Authority, Kuwait Al-Youm: Kuwait, 2017.

MACROINVERTEBRATES AS BIOINDICATORS OF WATER QUALITY IN THE METROPOLITAN REGION OF PANAMA

MADELINE CONCEPCIÓN[1], LIZ CANO[1], QUIRIATJARYN ORTEGA[2], ANDRES FRAIZ[3], ADRIAN RAMOS-MERCHANTE[4], ARTURO DOMINICI-AROSEMENA[5], EDGAR ARAÚZ[1], EDGARDO MUÑOZ[1] & ARAMIS AVERZA[1]
[1]University of Panama, Panama
[2]Polytechnic University of Valencia, Spain
[3]Foundation Wetland International, Panama
[4]University of Huelva, Spain
[5]International Maritime University of Panama, Panama

ABSTRACT

This research is part of the work on the environmental impact of multi-stressors on the aquatic ecosystems of the metropolitan area of Panama City, which is aligned to the strategy of Resilience of Panama City, about a plan of actions within the economic, social, and environmental field. This paper presents a study of bioindicators of water quality using the macroinvertebrate communities present along (high, middle, and low points) of three different basins of the Pacora, Matasnillo, and Juan Díaz rivers at different times of the year (dry, rainy, and transitional seasons). Biological monitoring carried out in sections of 100 m were taken and samples collected using a D-Net network of 500 µm, the organisms were examined in the laboratory with the support of specialists from the Gorgas Memorial Institute for Health Studies. Based on the compilation and review of the material, we were able to list a total of 2,704 individual freshwater macroinvertebrates. According to the BMWP/PAN results, it is observed that the highest levels of pollution are found in the lower parts of each tributary. The Matasnillo River has the highest pollution range with a value of <19, while the Pacora River has the lowest range. The results are important information about the functional groups of aquatic insects and provide relevant data for monitoring and conserving rivers, streams, and others. Due to the difficulties in educating the population on how to care for our aquatic ecosystems, it is important to diagnose water quality using the presence or absence of aquatic invertebrates.

Keywords: multi-stressors, resilience, bioindicators, macroinvertebrates, basin, BMW-PAN, freshwater, water quality, pollution.

1 INTRODUCTION

Aquatic macroinvertebrates take a huge part of the aquatic diversity, so they are often the main animal component of a lotic ecosystem [1] These provide the following characteristics. They are abundant, widely distributed, and easy to collect, easy to identify, if compared to other minority groups are mostly sedentary and reflect local conditions, have long life cycles, are visible to the naked eye, can be cultivated in the laboratory, very little genetically, respond quickly to environmental stressors [2]. In Panama, watersheds have undergone a great transformation because they are a source of natural resources, receive both agricultural and industrial waste, and are also used for recreation. On a regional scale, the basin of Pacora (146), Matasnillo (8142) and Juan Diaz (144) rivers, especially basin 146, which has a high degree of transformation since it has been modified to generate water supply to remote populations in the region, in addition to sustaining social activities, cultivation, industrial and economic use, extraction of stone and sand from the river, tourist uses, among others [3]. Therefore, it is necessary to evaluate the degree of deterioration of different basins in an integrated manner to allow a more complete analysis, provide a better criterion for the level of affectation of the basin, and provide future solutions.

www.witpress.com, ISSN 1743-3541 (on-line)
doi:10.2495/WMEI220041

2 MATERIALS AND METHODS

2.1 Areas of study

2.1.1 Basin 144 – Juan Díaz River

The sampling stations were in the upper (JD001), middle (JD002), and lower (JD003) basins of the Juan Díaz River at different times, located on the Pacific slope within the province of Panama (Table 1). Between the coordinates 9°00′ and 9°13′ and north latitude 79°18′ and 79°34′ west longitude [4]. It is located southeast of Panama City.

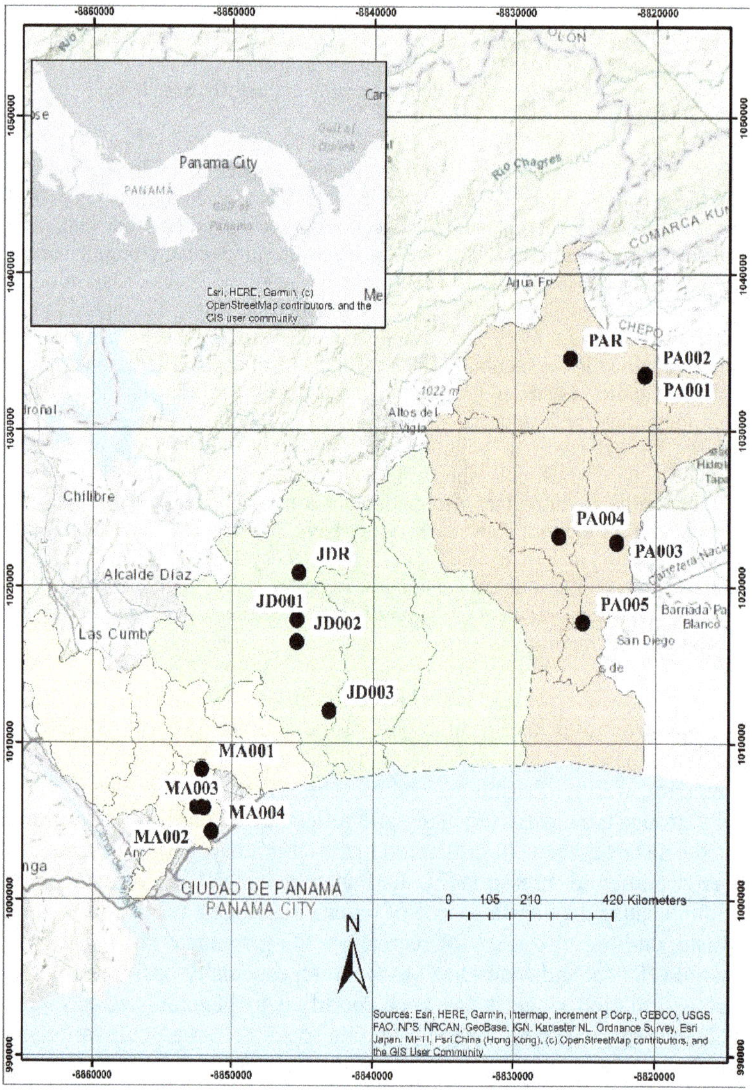

Figure 1: Representative map of the sampling stations of the Juan Díaz, Pacora, and Matasnillo rivers.

Table 1: Data from the rivers Juan Diaz, Pacora, and Matasnillo.

Data from the Juan Diaz, Pacora, and Matasnillo rivers				
Basin	Basin number	Season	X	Y
Juan Diaz	No. 144	JRD	669349	1003785
		JD001	669212	1006831
		JD002	669211	1005437
		JD003	671529	1001140
Pacora	No. 146	PAR	688289	1023313
		PA001	693513.5	1022331.2
		PA002	693479.6	1022271.6
		PA003	691615	1011796
		PA004	687568	1012143
		PA005	689299	1006779
Matasnillo	No. 142	MA001	662638	997397
		MA002	662314	995064
		MA003	662823	995028
		MA004	663364	993524

It has two types of climates, temperate tropical savannah and humid tropical. The basin has an average annual rainfall of 2,466 mm. The vegetation is represented by three ecological communities: area covered by forest (the secondary forest little intervened and secondary forest very intervened), stubble area, and wetlands (mangroves). Of the total area, 8% is used for urban uses, such as low-density residential, commerce and services, institutions, industry, transportation and communication, recreation, and green areas [4].

2.1.2 Basin 146 – Pacora River

The sampling stations were in the upper (PA001, PA002), middle (PA003, PA004), and low (PA005) basins of the Pacora River at different times of the year. This river belongs to the province of Panama, Panama District, located on the pacific slope, between the coordinates 8°00′ and 8°20′ north latitude and 79°15′ and 79°30′ west longitude [3]. On its way, it collects water from important tributaries, such as the Tataré, Utivé, Calobre, and Indio rivers [5].

It has two types of climates, tropical temperate savannah which represents 60% of the total area of the basin, and tropical humid climate that represents 40% [4]. There are diverse ecosystems represented in three life zones: very humid premontane forest, premontane rainforest, and tropical rainforest [5].

2.1.3 Basin 142 – Matasnillo River

The samplings stations were in the upper (MA001), middle media (MA002, MA003), and lower (MA004) basins of the Matasnillo River at different times of the year. It is the only river that rises and flows into the Panamanian capital, its length is six km.

It has no main course, and its course is the result of a series of streams that extend in the district of Bethania (northernmost point), Pueblo Nuevo, Bella Vista, and San Francisco. It is located on the pacific slope, within the province of Panama, and occupies an area of

383 km^2, representing 0.51% of the national territory. Its geographical coordinates are 8°50′ and 9°05′ north latitude and 79°30′ and 79°40′ west longitude [5].

2.2 Field and laboratory work

Qualitative collections were made in three different seasons (rainy, dry, and transition) in each season a sampling effort of 30 minutes was made. To do this, the samples were collected using 500 µm D-net in vertical position taking it by the highest part of the handle (grip) and placing it against the river current in contact with the bottom of the surface of the river to be sampled, depending on the case of the river, it will be sampled once in each of the groups of microhabitats that we identified trying that the sample does not spread outside the net. The samples were then placed in labelled bottles with screw caps and preserved in 96% alcohol. These were transferred to the laboratory where they were processed. When we settled in, we examined the collected material thoroughly, under optical equipment, for the process of cleaning and separation into larger groups, with 96% alcohol. And we proceeded to make the identification at the taxonomic level of the family. The data of the identified organism were grouped by season (rainy, dry, and transition), so we obtained a few families, genera, and individuals for each sampling season. Water quality was determined at the 14 sampling points in the Pacora, Matasnillo and Juan Díaz rivers, using BMWP/PAN (Biological Monitoring Working Party/Panama) [6].

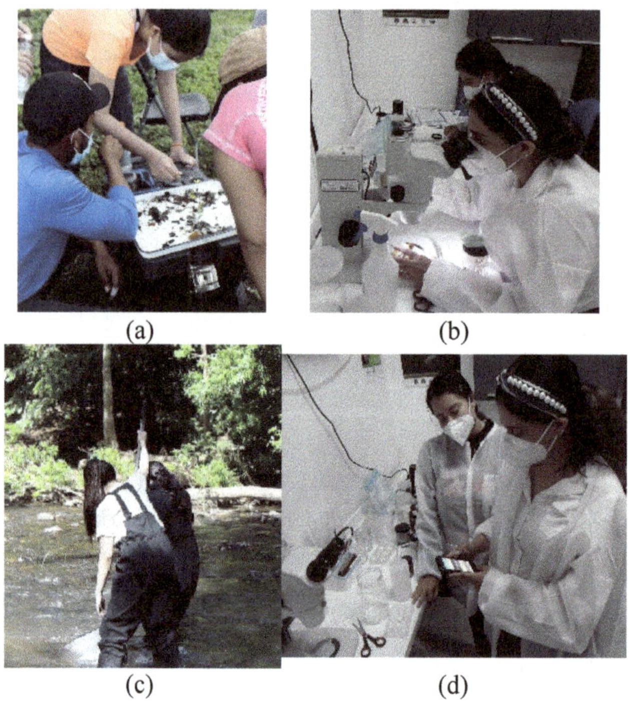

Figure 2: (a) Separation of macroinvertebrates in the field; (b) Identification at the taxonomic level of families; (c) Using red of collect in vertical position taking it by the highest part of the handle (grip) and placing it against the river current in contact with the bottom of the surface of the river to be sampled; and (d) Determining the BMWP.

3 RESULTS

3.1 Abundance and wealth of macroinvertebrates in the basin 144 – Río Juan Díaz

A total of 2,704 individuals of freshwater macroinvertebrates were collected. Ninety-three families distributed in eight orders were recorded, being the Ephemeroptera (309 individuals) and Diptera (261 individuals) the most abundant order.

The most abundant families were Chironomidae (Diptera) with 30.14% of the total reported, Leptohyphidae (Ephemeroptera) with 18.29%, Baetidae (Ephemeroptera) with 17.43%, Leptophlebiidae (Ephemeroptera) con 8.43%, Glossosomatidae (Trichoptera) with 7.14%, Hydropsychidae (Trichoptera) with 4.57%, Simuliidae (Diptera) with 3.43%, Psephenidae (Coleoptera) with 2.14% and Psychodidae (Diptera) with 1.71%. These families represented 70.01% of the total reported of this basin.

3.1.1 Biological water quality

"Indirectly, rainfall influences the behaviour of the chemical parameters of the water" [2]. Directly, insects are dragged by currents and deposited in different parts of the river channel, influencing their distribution [7]. Therefore, the fact that our data show few individuals during the rainy season compared to the dry season and the transition period may be due to these relationships. Also, it is worth noting the strong currents and rocky substrates, which complicate sampling a bit at the time of kicking.

Table 2: Biological water quality in the 144 river basin (rainy season).

Biological water quality in the 144 river basin			
	JD001LL	JD002LL	JD003LL
Family richness	7	15	9
Numbers of individuals	40	160	115
BMWP/PAN score	23	50	27
Significance	Highly polluted	Polluted	Highly polluted

Table 3: Biological water quality in the 144 river basin (dry season).

Biological water quality in the 144 river basin				
	JDR	JD001SE	JD002SE	JD003SE
Family richness	35	34	27	7
Numbers of individuals	442	438	546	241
BMWP/PAN score	155	151	103	15
Significance	Excellent	Excellent	Good	Extremely polluted

Table 4: Biological water quality in the 144 river basin (transition season).

Biological water quality in the 144 river basin			
	JD001TR	JD002TR	JD003TR
Family richness	33	28	6
Numbers of individuals	288	363	74
BMWP/PAN score	76	86	13
Significance	Regular	Good	Extremely polluted

3.2 Abundance and richness of macroinvertebrates in the basin 146 – Río Pacora

A total of 2,143 individuals of freshwater macroinvertebrates were collected, divided into three classes, being Insecta (2,112 individuals) the most represented. Only 31 individuals were part of other taxonomic groups (Gasteropoda and Malacostraca). Twenty-eight families distributed in 10 orders were recorded, being the orders Ephemeroptera (1165 individuals), Trichopteran (517 individuals), and Diptera (203 individuals) the most abundant. The most abundant families were Leptophlebiidae (Ephemeroptera) with 18.11%, Hydropsychidae (Trichoptera) with 10.73% Philopotamidae (Trichoptera) witch 9.38%, Chironomidae (Diptera) with 6.86%, Glossosomatidae (Trichoptera) with 5.97%, Baetidae (Ephemeroptera) with 4.53%, Psephenidae (Coleoptera) with 1.91% and Coenagrionidae (Odonata) with 1.59%. These families represented 90.71% of the reported for this basin.

3.2.1 Biological water quality

Table 5: Biological water quality in the 146 river basin (rainy season).

Biological water quality in the 146 river basin					
	PA001LL	PA002LL	PA003LL	PA004LL	PA005LL
Family richness	19	18	12	17	14
Numbers of individuals	170	113	140	165	239
BMWP/PAN score	62	54	41	58	49
Significance	Regular	Polluted	Polluted	Polluted	Polluted

Table 6: Biological water quality in the 146 river basin (dry season).

Biological water quality in the 146 river basin						
	PAR	PA001SE	PA002SE	PA003SE	PA004SE	PA005SE
Family richness	35	38	26	18	24	16
Numbers of individuals	465	389	318	220	229	365
BMWP/PAN score	154	171	103	76	88	53
Significance	Excellent	Excellent	Good	Regular	Good	Polluted

Table 7: Biological water quality in the 146 river basin (transition season).

Biological water quality in the 146 river basin					
	PA001TR	PA002TR	PA003TR	PA004TR	PA005TR
Family richness	34	36	41	23	12
Numbers of individuals	282	310	412	106	150
BMWP/PAN score	88	101	65	61	30
Significance	Good	Good	Regular	Regular	Highly polluted

3.3 Abundance and richness of macroinvertebrates in the basin 142 – Matasnillo River

A total of 434 individuals of freshwater macroinvertebrates were collected. In which, 21 families were recorded, being Thiaridae (309 individuals) and Ceratopogonidae (261 individuals) the most abundant.

The most abundant families were: Thiaridae (Mesogastropoda) with 39.17% of the total reported, Ceratopogonidae (Diptera) with 21.66%, Chironomidae (Diptera) with 11.29%, Tubificidae (Haplotaxida) with 9.91%, Tubificidae (Haplotaxida) with 5.53%, Syrphidae (Diptera) with 3.00%, and Tipulidae (Diptera) with 1.38%. These families represented 91.94% of the total reported for this basin.

3.3.1 Biological water quality

Table 8: Biological water quality in the 142 river basin (rainy season).

Biological water quality in the 142 river basin				
	MA001LL	MA002LL	MA003LL	MA004LL
Family richness	4	2	2	2
Numbers of individuals	83	9	4	9
BMWP/PAN score	14	3	3	3
Significance	Extremely polluted	Extremely polluted	Extremely polluted	Extremely polluted

Table 9: Biological water quality in the 142 river basin (dry season).

Biological water quality in the 142 river basin				
	MA001SE	MA002SE	MA003SE	MA004SE
Family richness	8	8	2	4
Numbers of individuals	88	65	2	58
BMWP/PAN score	22	17	3	5
Significance	Highly polluted	Extremely polluted	Extremely polluted	Extremely polluted

Table 10: Biological water quality in the 142-river basin (transition season).

Biological water quality in the 142 river basin				
	MA001TR	MA002TR	MA003TR	MA004TR
Family richness	9	2	2	8
Numbers of individuals	37	3	8	68
BMWP/PAN score	20	3	4	11
Significance	Highly polluted	Extremely polluted	Extremely polluted	Extremely polluted

4 DISCUSSION

4.1 Biotic water quality index BMWP/PAN (Biological Monitoring Working Party)

Our study shows that watersheds 142 and 144 show serious deterioration. In the first tributary, four of the sites evaluated showed low richness of macroinvertebrate families and with a dominance of those that are tolerant to pollution. Stations MA001SE and MA001TR showed the highest richness of families and obtained values of 20 and 22, which corresponds to the category of highly polluted water quality. Meanwhile, in the second basin, three of the sites evaluated showed low richness of families in different seasons of the year, and with dominance to those that are tolerant to pollution. This is mainly the case of stations JD003LLL, JD003SE and JD001TR, which registered the lowest richness of families and obtained scores between 13 and 27, corresponding to extremely polluted waters. The second station with the highest richness was JD001SE, which corresponds to the category of excellent water quality. However, we can observe that as the rainy season approaches it shows a decrease in the number of taxa found. This is due to the fact that "the abundance of insects is mainly observed when rainfall is low and decreases when it increases" [7]. Similar results to those found in this study were documented in the Garachiné river [8], who also found a low abundance of EPT during periods of high rainfall "probably caused by an increase in water flow, which produces a consequent drift effect of individuals, the removal of insects and a reduction in their local abundance" [9].

It is important to highlight the small tributaries that cross the urbanizations and shopping centres, which have become recipients of all kinds of waste, either by runoff or simply because they are thrown directly into the tributaries due to a lack of environmental culture.

The JDR station was the one that presented the highest richness of families and obtained a value of 155, which corresponds to the category of Excellent water quality (Table 6); however, this is a reference point with one of the highest altitudes in the basin. This shows that the basin is being affected by the different activities carried out by people and will show greater deterioration every day if necessary, measure are not taken to mitigate the problem.

In the case of watershed 146, the quality of the watershed was acceptable despite the activities carried out along the watershed. However, of our six sampling points, three showed low richness of macroinvertebrate families in different seasons of the year. And with this we confirm once again that: "Rainfall indirectly influences the behavior of the chemical parameters of the water" [11]. In a direct way the insects are dragged by the currents and deposited in different parts of the riverbed influencing their distribution [7]. Therefore, the fact that our data have few individuals during the rainy season compared to the dry and transitional seasons is due to these relationships. This is the case of stations PA003, PA004 and PA005; which registered the lowest richness of families and obtained low scores, which on average corresponds to very polluted waters. With this, we can deduce that the signs of deterioration in their ecological quality are mainly in the middle and lower reaches of the river, which correspond to 50 and 100 m alongside the river. PA005 is the most affected at both times of the year. It should be noted that in this sampling station is located the water intake that supplies the water treatment plant of Pacora and where economic activities are developed. "Therefore, it is necessary to make interventions on the issue of the elimination of vegetation cover on the riverbanks, the extraction of stone material, the management of wastewater and solid waste" [6]. Similar results to those found in this study, were documented in Cornejo [6] precisely in this same watershed. Where, she obtained a BMWPA/PAN index of 62 which corresponds to a regular quality. While we obtained an index of 49 during the rainy season, 53 during the dry season and 30 during the transition period, which means that this point is being affected on a large scale and we see quite marked deterioration in a short period of time.

4.2 Physical and chemical variables

4.2.1 Basin 144 – Juan Díaz River

In the dry season, on average, the levels of dissolved oxygen (8.17 mg/L), pH (7.78), and conductivity (199.9 µS/cm) remained at normal levels in all sampling stations (Table 11), by the values established by the primary environmental quality standards and quality levels for inland waters for recreational use with and without direct contact [4]. During the rainy season, levels were below the established limits for dissolved oxygen (5.56 mg/L) and pH (7.05). Conductivity values remained within the accepted range (112.38 µS/cm). Normal values of dissolved oxygen ($\geq$ 5 mg/L) and pH (6.0–9.0) are important in aquatic ecosystems, as they are related to the metabolic processes of aerobic organisms and biological productivity, respectively [1]. In sampling stations JD001 and JD003, where the highest conductivity values were obtained, possibly influenced by wastewater discharge, lower species diversity was observed, compared to that of station JD002, in the rainy season. Although the influence of other factors, such as altitude and/or water temperature, is not ruled out, it is known that conductivity affects species diversity; as conductivity increases, diversity decreases [10].

4.2.2 Basin 146 – Pacora River

In the dry season, on average, the levels of dissolved oxygen (8.31 mg/L), pH (7.97), and conductivity (182.06 µS/cm) remained at normal levels at all sampling stations (Table 12), by the values established by the primary environmental quality standards and quality levels for inland waters for recreational use with and without direct contact [4]. During the rainy season, levels were below the established limits for dissolved oxygen (7.51 mg/L) and pH (7.52).

In the dry season, on average, dissolved oxygen levels (8.17 mg/L), pH (7.78) and conductivity (199.9 µS/cm) remained at normal levels at all sampling stations (Table 11), by the values established by the primary environmental quality standards and quality levels for inland waters for recreational use with and without direct contact [4]. Whereas, in the rainy season, the levels were lower than the limits established for dissolved oxygen (5.56 mg/L) and pH (7.05). Conductivity values were maintained within the accepted range (112.38 µS/cm).

Table 11: Values of the BMWP/PAN biotic index for water quality and physicochemical variables at each sampling station during the rainy (LL), dry (SE), and transition (TR) seasons in the Juan Diaz River basin. Normal ranges: LDO: 7.0–8.0 mg/L, pH: 6.0–9.0, conductivity: 100–200 µS/cm [7], [10].

Physical and chemical variables of the 144 Juan Diaz river basin in the rainy, dry and transition periods					
Season	Variable	BMWP/PAN	LDO (mg/L)	pH	Conductivity (µS/cm)
JD001	JDR	155	8.65	7.97	133.6
	LL	23	8.7	7.07	134.5
	SE	151	8.35	8.05	140.5
	TR	76	8.27	7.65	116.8
JD002	LL	50	7.88	7.2	370
	SE	103	7.9	7.75	135.6
	TR	86	7.34	7.58	131.5
JD003	LL	27	0.1	6.9	138.8
	SE	15	7.81	7.35	323.6
	TR	13	7.55	6.68	316.8
Average	LL	33.33	5.56	7.05	214.43
	SE	89.66	8.2	7.78	199.9
	TR	58.33	7.72	7.3	188.36

Table 12: Values of the BMWP/PAN biotic index for water quality and physicochemical variables at each sampling station during the rainy (LL), dry (SE) and transition (TR) seasons in the Pacora River basin. Normal ranges: LDO: 7.0–8.0 mg/L, pH: 6.0–9.0, conductivity: 100–200 µS/cm [7], [10].

Physical and chemical variables of the 146 Pacora river basin in the rainy, dry and transition periods					
	Variable	BMWP/PAN	LDO (mg/L)	pH	Conductivity (µS/cm)
PA001	PAR	154	9.41	8.07	159.1
	LL	62	9.15	7.11	124.4
	SE	171	7.76	8.5	177.4
	TR	88	7.68	7.91	170.6
PA002	LL	54	9.34	7.5	135.3
	SE	103	9.21	7.93	165.9
	TR	101	8.08	8.03	161.3
PA003	LL	41	6.16	7.6	154.5
	SE	76	8.19	8.14	198.5
	TR	65	7.93	7.82	178.6
PA004	LL	58	6.11	7.66	139.8
	SE	88	7.66	7.62	178.2
	TR	61	6.89	7.73	179.6
PA005	LL	49	6.8	7.77	169.2
	SE	53	8.73	7.68	190.3
	TR	30	8.61	7.58	207.1
Average	LL	52.8	7.51	7.52	144.64
	SE	98.2	8.5	7.99	182.06
	TR	69	7.83	7.81	179.44

Table 13: Values of the BMWP/PAN biotic index for water quality and physicochemical variables at each sampling station during the rainy (LL), dry (SE), and transition (TR) seasons in the Pacora River basin. Normal ranges: LDO: 7.0–8.0 mg/L, pH: 6.0–9.0, conductivity: 100–200 µS/cm [7], [10].

Physical and chemical variables of the 142 Matasnillo river basin in the rainy, dry and transition periods					
	Variable	BMWP/PAN	LDO (mg/L)	pH	Conductivity (µS/cm)
MA001	LL	14	3.8	7.77	302
	SE	22	3.61	7.36	380.4
	TR	20	3.83	7.08	384.1
MA002	LL	3	1.15	7.54	600
	SE	17	1.99	7.23	629
	TR	3	2.91	6.58	613
MA003	LL	3	0.9	6.81	600
	SE	3	2.78	7.37	546
	TR	4	2.44	7.4	510
MA004	LL	3	1	7.91	554
	SE	5	2.84	7.84	568
	TR	11	1.2	7.08	555
Average	LL	5.75	1.71	7.5	514
	SE	11.75	2.8	7.45	530.85
	TR	9.5	2.59	7.03	515.52

5 CONCLUSION

In the first place, we were able to evaluate the quality of the water with macroinvertebrates in an altitudinal gradient comparing with the biological results and previous studies of water quality, we were able to classify the most representative families of benthic macroinvertebrates present in the Juan Diaz, Matasnillo and Pacora rivers.

Of the three watersheds treated and according to the BMWP, we were able to observe that the Matasnillo River has the highest range of contamination with a value of <19 and the Pacora River has the lowest range.

The physical and chemical variables showed that during the rainy season the levels were low, by the values established by the primary environmental quality standards and quality levels for inland waters for recreational use with and without direct contact, influencing the heterogeneity of the aquatic macroinvertebrates found in each sampling station. We consider that climate was an important factor in our sampling. Since it, directly and indirectly, influenced the abundance of insects in the river.

In the Matasnillo River, we obtained a total of 434 individuals distributed in 21 families, in the Juan Díaz River 2707 individuals were distributed in 48 families, and in the Pacora River 4073 individuals were distributed in 54 families. The total number of individuals in our sampling basins was 7214. Of which, the highest to the lowest abundance of families we have in the Pacora River; and the most predominant Atyidae. In the Juan Díaz River, Atyidae, and the Matasnillo River, the Thiaridae family, therefore, it is corroborated that in the higher areas less contamination can be observed.

ACKNOWLEDGEMENTS

We thank the National Secretariat of Science, Technology, and IOMA Project 19-008: Environmental impact of multistressors on aquatic ecosystems of the metropolitan area of Panama, Innovation (SENACYT), Quiriatjaryn Ortega, the Wetland International Foundation, our advisors Aramis Averza, Edgar Araúz, and Edgardo Muñoz, and researchers Aydeé Cornejo and Carlos Nieto of the Gorgas Memorial Center for their support and expanding our knowledge. and to our colleague Adolfo Alba for supporting us and helping us in the grammar of the paper.

REFERENCES

[1] Nieves, G.E., Rosas R.K. & Hornedo, M., Biodiversity of aquatic insects associated with the basin of the Rio Grande de Manatí. Department of Natural and Environmental Resources, Puerto Rico, 2010. http://www.drna.gobierno.pr/oficinas/saux/secretaria-auxiliar-de-planificacion-integral/planagua/proyecto-rios-patrimoniales/estudio-debiodiversidad-de-insects-acuaticos.pdf. Accessed on: 25 Jul. 2012.

[2] Roldan Pérez, G., Macroinvertebrates and their value as indicators of water quality. *Rev. Acad. Colomb. Cienc.*, **23**(88), pp. 375–387, 1999.

[3] Regional Program for the Reduction of Vulnerability and Environmental Degradation, Formulation of the Strategic Plan for the Integrated Management of the Pacora River Basin, Panama, Republic of Panama, 121 pp., 2008.

[4] ANAM (National Environmental Authority), Report on monitoring water quality in the watersheds of Panama, Compendium of Results, Years 2006–2007, Republic of Panama, 2009.

[5] ANAM (National Environmental Authority), Report on monitoring water quality in the watersheds of Panama, Compendium of Results, Years 2009–2012, Republic of Panama, 2013.

[6] Cornejo, R.A., Aquatic macroinvertebrates bioindicators of water quality in Panama: BMWP/PAN index proposal. Special meeting: Freshwater macroinvertebrates in Mesoamerica (MADMESO), Villahermosa, Tabasco, Mexico, 2010.

[7] Araúz, B., Amores, B. & Medianero, E., Diversity of distribution of aquatic insects along the bed of the Chico River, province of Chiriquí, Republic of Panama. *Scientia*, **15**(1), pp. 27–45, 2000.

[8] Del C. Guinard, J., Rivers, T. & Bernal Vega, J.A., Diversity and abundance of aquatic macroinvertebrates and water quality of the upper and lower basins of the Gariché River, Chiriquí Province. *Panama Management and Environment*, **16**(2), pp. 61–70, 2013.

[9] Flecker, A.S. & Feifarek, B., Disturbance and the temporal variability of invertebrate assemblages in two Andean streams. *Freshwater Biology*, **31**, pp. 131–142, 1994.

[10] Roldán, G., Bioindication of water quality in Colombia. Use of the BMWP/Col.Editorial Universidad de Antioquia method. Medellin, Colombia, 170 pp., 2003.

[11] Rincón, H.M.E., Comunidad de insectos acuáticos de la quebrada Mamarramos (Boyacá-Colombia). *Revista Colombiana de Entomología*, **28**(1), pp. 101–108, 2002.

SECTION 2
CIRCULAR ECONOMY

AGRO-WASTE ASHES AS A FEEDER FOR THE SYNTHESIS OF SIO_2 NANOPARTICLES FOR ROAD CONSTRUCTION

MARTIN ALUGA & KAMBOLE CHEWE
Civil Engineering and Construction Department, School of Graduate Studies, Copperbelt University, Zambia

ABSTRACT

The global rapid population growth is leading to accelerated food production thus accelerating the generation of agricultural wastes (AWs). Among the AWs management strategies is the valorisation of the AWs into useful products such as the production of silica nanoparticles (SNPs) which are useful in the road construction sector. Production of useful SNPs from AWs for the road construction industry is considered sustainable road construction, as not only does it reduces the carbon dioxide footprint from similar materials such as cement, but it is also cheap and readily available to low-income countries such as Kenya and Zambia. SNPs are of great importance in the production of high-strength materials such as stabilized soil and asphalt of significant strength than the conventional stabilised soil/asphalt with ordinary Portland cement. In this review, recent progress on AWs corresponding ashes (AWAs) with significant silica is studied. The production of SNPs from this AWAs is examined in detail concerning their synthesis methods for high purity SNPs, structure, phase, morphology, sizes, and their application in soil and asphalt stabilisation in road construction. The use of AWs to produce SNPs for use in the stabilisation of soils, bitumen, and asphalt for roads construction is promising, however, there is a need for further research on the large-scale production of SNPs from AWAs.

Keywords: biomass, construction, nanomaterials, nanotechnology, ordinary Portland cement (OPC), silica nanoparticles.

1 INTRODUCTION

Agricultural wastes are mainly of four types as shown in Fig. 1. They are composed of crop, food, industrial processing, and livestock wastes. They are generated after the useful part of the crop is extracted or while rearing livestock. Just as any other wastes, AWs need to be managed properly for a hygienic environment through different waste management strategies. These strategies include utilising the AWs in the production of novel products ass value-addition to the agricultural economy. These AWs are burned to produce AWs Ashes (AWAs).

Some of the novel products that can be produced from AWAs are SNPs. SNPs are either produced by bottom-up or up-bottom methods using minerals that are non-renewable such as silica sand. There is a need to minimize the use of non-renewable sources to produce SNPs. This has led to research on using AWAs to produce SNPs that are considered as green. It results in AWs management while enhancing value addition in agriculture. These SNPs can used as pozzolanic additives in civil engineering projects for strengthening concrete, and the production of plastics and rubber materials. This results in sustainable construction.

In this review, recent progress on AWAs with significant silica is studied. The production of SNPs from AWAs is examined in detail about their synthesis methods for high purity SNPs, sizes, and their application in soil, bitumen, and asphalt stabilisation in road construction sector.

2 AGRO-WASTES CONTAINING SILICA

Table 1 shows some of the AWAs containing silica as characterised by x-ray fluorescence spectroscopy (XRF) from recent studies. Rice straw and husk ashes as crop waste and

www.witpress.com, ISSN 1743-3541 (on-line)
doi:10.2495/WMEI220051

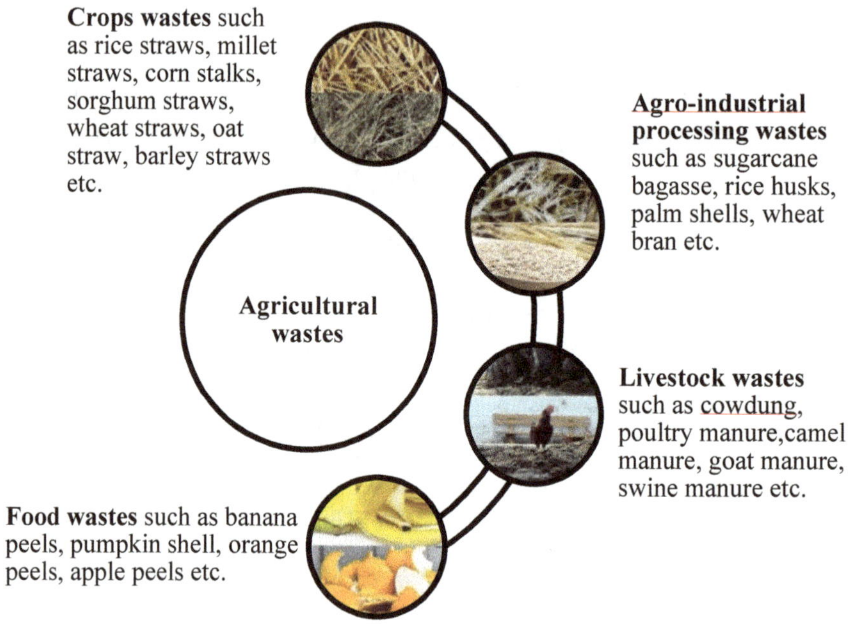

Figure 1: Types of agricultural wastes.

Table 1: Agricultural wastes ashes containing silica as characterised by x-ray fluorescence spectroscopy (XRF) from different studies.

Biomass ash	Silica (SiO_2) (%)	Reference
Rice husk ash	72.2–99.50	[1]–[6]
Areca sheath ash	78.00	[7]
Sugar cane bagasse ash	40.46–98.40	[8]–[16]
Palm leaf ash	53.11–82.15	[17], [18]
Corn cob ash	64.80	[19]
Barley straw ash	55.00–61.75	[20], [21]
Bamboo leaf ash	72.81	[22]
Millet husk ash	69.40	[23]
Wheat straw ash	67.83	[23]
Wood ash	7.75–68.51	[24]–[27]
Sorghum husk ash	70.48	[28]
Paddy waste ash	84.78	[29]
Ficus leaf ash	81.45	[30]
Palm oil fuel ash	61.68	[31]
Sunflower ash	50.36	[32]
Anacardium ash	65.02	[33]
Cowdung ash	57.33–61.00	[34], [35]

industrial processing waste respectively are the main studied AWAs as they contain a lot of silica. Cow dung ash and areca sheath ash represents the livestock and food wastes respectively. Silica is one of the main components of any material to be characterised as a pozzolanic material in road construction or any general construction works. Alongside silica, are the alumina and ferrous oxides.

3 SYNTHESIS OF SILICA NANOPARTICLES FROM AGRICULTURAL WASTES ASHES

Table 2 shows the studies that have been undertaken on the production of SNPs from the AWAs. RHA is the main studied material among the AWAs. Several reviews have been done on the production of SNPs from AWAs [36], [37], however, most of the reviews focus only on single crop residues. Agricultural wastes consist also of food and livestock wastes. Some studies have also focused on specific crops such as rice husk ash and straws ashes to produce novel products [38], [39] This review brings together all together these studies to make some general conclusions from the existing data.

The sol-gel method is the most preferred synthesis method. Few studies exist on other methods such as sonochemistry. This is the case because the sol-gel process involves producing solid materials from small molecules [49]. UV-visible spectroscopy (UV-Vis), FE-SEM energy-dispersive x-ray (EDX), x-ray diffraction (XRD), and FTIR, are some of the main spectroscopy techniques used for characterisation of the produced SNPs. Most of the SNPs produced from the AWAs are amorphous.

4 APPLICATION OF AMORPHOUS SILICA NANOPARTICLES

This research reviews the use of amorphous SNPs in road construction, especially on soil stabilisation, and the use of SNPs additives in bitumen or asphalt as detailed in Table 3. Few studies are published on the use of SNPs for soil stabilisation for road construction [50]. In addition, studies on the use of SNPs as additives in bitumen and asphalt exists though limited [51].

The review indicates that residual soil can be improved by over 200% in terms of compressive strength when 15 nm-SNPs are added [50]. The penetration grade and viscosity grade of bitumen and asphalt are also improved when SNPs are used. This is mainly on aging and fatigue resistance as detailed in Table 3. Bitumen and asphalt are stabilised against rutting, cracking, and temperature susceptibility. The elasticity of bitumen and asphalt increases, with more resistance against the flowing, and decreased the fatigue life.

5 CONCLUSION

In the current research, recent progress on AWAs with significant silica has been listed. RHA and SCBA are the main AWAs being studied with high silica content. The production of SNPs from this AWAs is examined in detail with regard to the synthesis methods for high purity silica nanoparticles, characterisation, and its application in soil and asphalt stabilisation in road construction. The study shows that the sol-gel method is the most preferred synthesis method for SNPs from AWAs. UV-Vis, FTIR, FESEM, EDX, and XRD are some of the main techniques used for the characterisation of the produced SNPs. The results show that most of the SNPs produced from the AWAs are amorphous with sizes ranging from 5 to 75 nm. Though there exist latest few studies on the application of amorphous SNPs in road construction, they are promising with regard to improving the compressive strength of soil and the performance of bitumen and asphalt thus longer life span.

Table 2: Production of SNPs from some agro wastes.

Agro waste ash	Synthesis method	Characterisation and analysis	Characteristics	Ref.
Rice husk ash (RHA)	The sol-gel technique.	Ultraviolet-visible spectroscopy (UV-Vis), Fourier transform infrared (FTIR), field emission scanning electron microscope (FE-SEM), EDX, and XRD were used in this study.	Amorphous SNPs with sizes of 15–20 nm.	[40]
Palm shell ash (PSA)	The modified sol-gel technique.	XRD, SEM, EDX, FTIR.	Amorphous SNPs with sizes ranging from 50 nm to 98 nm. The surface area was 438 $m^2\ g^{-1}$.	[41]
Sugarcane bagasse ash (SCBA)	The Sol-gel polymeric method	X-ray fluorescence spectroscopy (XRF), SEM equipped with EDX and TEM were employed.	SNPs, are obtained as a nanostructured powder of about 10 nm in diameter with over 95% of purity.	[42]
Bamboo leaves ash	The Sol-gel method	UV-Vis and FTIR.	The FTIR spectrum confirms the presence of SNPs within the functional groups as shown in Fig. 2.	[43]
Banana peels ash	The sol-gel method	FTIR instrument.	The FTIR spectrum had functional groups ranging from 400 cm^{-1}. to 4,000 cm^{-1}. Confirming the presence of SNPs.	[44]
Rice husk	The chemical method	XRD, FE-SEM, EDAX, and TEM.	The SNPs obtained were between 3 nm to 10 nm and with the highest surface area of about $247m^2/g^{-1}$).	[45]
Rice husk ash	The sonochemical process	FE-SEM, EDX, TEM and FTIR.	SNPs obtained had sizes from 5 nm to 40 nm and the surface area from 176 m^2g^{-1} to 226 m^2g^{-1}.	[46]
RHA, BLA, SCBA, and groundnut shell ashes	The dissolution and precipitation process	FT-IR, XRD, and SEM-EDAX.	SNPs produced had a diameter ranging from 20 nm to 40 nm.	[47]

Table 3: Production of SNPs from some agricultural wastes.

Material	Size of NPS	Tests done	Performance	Ref.
Residual soil	15 nm and 80 nm	Atterberg limits, hydraulic conductivity, pH value, unconfined compressive strength (UCS) tests	The maximum strength improved from 550kPa to more than 1600kPa when SNPs of 15 nm were added to 8% cement	[50]
Bitumen-64-22 penetration grade (PG)	12 nm	Aging index, dynamic shear characterization, and direct tension test (DTT)	The addition of SNPs improved the bitumen in terms of rutting, cracking, and temperature susceptibility	[52]
Asphalt binder (PG 64–22)	12 nm	Accelerated aging of asphalt binder (AAAB), and DTT among other tests	SNPs improved the flowing and elasticity properties of asphalt	[53], [54]
60/70 PG asphalt binder	20–40 nm, 175–225 m^2/g BET surface area	Rotational viscosity test (RVT), frequency sweep (FS) test, multiple stress creep recovery (MSCR) test, linear amplitude sweep (LAS) test, four-point bending beam fatigue test (FBBF), bending beam rheometer (BBR) test	The addition of SNPs increased the elasticity of binders, with more resistance against the flowing, and decreased the fatigue life	[55]
No. 90 base asphalt binder	About 30 nm	BBR test, FS test, LAS test, MSCR test, temperature sweep (TS) test, rheological property tests, physical property tests, microstructure tests	The addition of SNPs, presented the greatest fatigue life and anti-crack property	[56]
Viscosity-graded (VG) asphalt binders	Under 200 nm	MSCR	SNPs addition in viscosity graded asphalts enhances its resistance to permanent deformation and fatigue	[57]
35/50 PG	70 nm	Fatigue, deformation, stiffness tests	SNPs modification presented the best global performance	[58]
60/70 PG bitumen	20–30 nm, 180–600 m^2/g surface area	Simulation of aging asphalt binders, FS test, fatigue testing	SNPs-modified binders had a better performance of fatigue behaviour compared to the control binders	[59]
Different asphalt models	7.5 nm	Molecular dynamics (MD) simulations	SNPs reduce erosion from chlorides	[60]
85/100 PG bitumen	20–30 nm	Four-point bending beam test	Samples containing 5% SNPs had a higher rate of self-healing	[61]
PG bitumen 60/70 and 85/100	20–30 nm	Compressive and tensile strength tests, modified Lottman indirect tension test	SNPs resulted in a significant increase in compressive strength	[62]
PG 64-22 bitumen	20–30 nm	MSCR test	The SNPs improved bitumen resistance to rutting	[63]
In-Organic silt	5 to 50 nm	California bearing ratio (CBR)	The SNPs lead to improvement of bearing strength	[64]
Loessial sediments	20–30 nm	Electrokinetic method	The SNPs performed more than the conventional lime by 2%	[65]

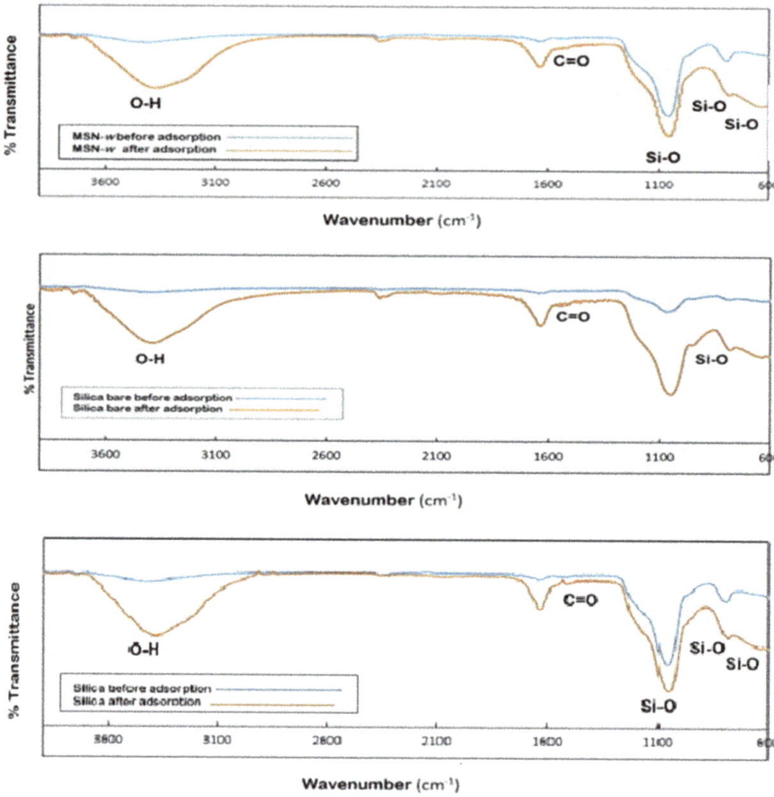

Figure 2: The FTIR spectrums of produced SNPs [43], [48].

6 CHALLENGES, OPPORTUNITIES, AND FUTURE RESEARCH

The use of AWAs to produce SNPs for use in the roads construction industry is promising though there is a need for further research on large-scale production. Large quantities of AWAs would need to be able to manufacture enough SNPs. Agricultural produce being perennial, there can be produced throughout the year. There is also need to undertake pilot real projects on the use of SNPs and its economic life cycle evaluated.

ACKNOWLEDGEMENTS

We appreciate the financial support of the Copperbelt University, Africa Center of Excellence for Sustainable Mining (CBU-ACESM) to work on this paper and attend the 11th ICWMEEISD Conference.

REFERENCES

[1] Abbass, M. & Singh, G., Impact strength of rice husk ash and basalt fibre based sustainable geopolymer concrete in rigid pavements. *Materials Today: Proceedings*, 2021.

[2] Bixapathi, G. & Saravanan, M., Strength and durability of concrete using rice husk ash as a partial replacement of cement. *Materials Today: Proceedings*, 2021.

[3] da Silva Nuernberg, N.B., Niero, D.F. & Bernardin, A.M., Valorization of rice husk ash and aluminum anodizing sludge as precursors for the synthesis of geopolymers. *Journal of Cleaner Production*, **298**, 126770, 2021.

[4] Jittin, V. & Bahurudeen, A., Evaluation of rheological and durability characteristics of sugarcane bagasse ash and rice husk ash based binary and ternary cementitious system. *Construction and Building Materials*, **317**, 125965, 2022.

[5] Mahdi, S.N., Babu, R.D.v., Hossiney, N. & Abdullah, M.M.A.B., Strength and durability properties of geopolymer paver blocks made with fly ash and brick kiln rice husk ash. *Case Studies in Construction Materials*, **16**, e00800, 2022.

[6] Wang, J., Fu, J., Song, W. & Zhang, Y., Effect of rice husk ash (RHA) dosage on pore structural and mechanical properties of cemented paste backfill. *Journal of Materials Research and Technology*, **17**, pp. 840–851, 2022.

[7] Kulkarni, P.P. & Siddeswarappa, B., A study on microstructure and mechanical behaviour of AA6063 metal matrix composite reinforced with areca sheath ash (ASA) and rice husk ash (RHA). *Materials Today: Proceedings*, 2021.

[8] Dang, L.C., Khabbaz, H. & Ni, B.J., Improving engineering characteristics of expansive soils using industry waste as a sustainable application for reuse of bagasse ash. *Transportation Geotechnics*, **31**, 100637, 2021.

[9] Dinesh Kumar, P.K., Rajendran, B., Subeetsha, B. & Swetha, N., Study on tensile, flexural and impact strength of GLARE reinforced with sugarcane bagasse ash microparticles. *Materials Today: Proceedings*, **47**, pp. 1059–1064, 2021.

[10] de Siqueira, A.A. & Cordeiro, G.C., Properties of binary and ternary mixes of cement, sugarcane bagasse ash and limestone. *Construction and Building Materials*, **317**, 126150, 2022.

[11] Nurwahid, I.H., Dimonti, L.C.C., Dwiatmoko, A.A., Ha, J.M. & Yunarti, R.T., Investigation on SiO_2 derived from sugarcane bagasse ash and pumice stone as a catalyst support for silver metal in the 4-nitrophenol reduction reaction. *Inorganic Chemistry Communications*, **135**, 109098, 2022.

[12] Salhotra, S., Khitoliya, R.K. & Kumar Arora, S., Assessing the enhanced concrete-properties induced by sugarcane bagasse ash-coated PET-fibers. *Materials Today: Proceedings*, **48**, pp. 994–1000, 2022.

[13] Sharma, T. & Singh, S., Influence of marble dust, bagasse ash and paddy straw fibers on the density and optimum water content of unfired soil block. *Materials Today: Proceedings*, **51**, pp. 965–971, 2022.

[14] Subedi, S., Arce, G.A., Hassan, M.M., Barbato, M., Mohammad, L.N. & Rupnow, T., Feasibility of ECC with high contents of post-processed bagasse ash as partial cement replacement. *Construction and Building Materials*, **319**, 126023, 2022.

[15] Wagh, M. & Waghe, U.P., Development of self-compacting concrete blended with sugarcane bagasse ash. *Materials Today: Proceedings*, 2022.

[16] Wu, N., Ji, T., Huang, P., Fu, T., Zheng, X. & Xu, Q., Use of sugar cane bagasse ash in ultra-high performance concrete (UHPC) as cement replacement. *Construction and Building Materials*, **317**, 125881, 2022.

[17] Amin, M., Zeyad, A.M., Tayeh, B.A. & Saad Agwa, I., Effects of nano cotton stalk and palm leaf ashes on ultrahigh-performance concrete properties incorporating recycled concrete aggregates. *Construction and Building Materials*, **302**, 124196, 2021.

[18] Yousefi, M., Khandestani, R. & Gharaei-Moghaddam, N., Flexural behavior of reinforced concrete beams made of normal and polypropylene fiber-reinforced concrete containing date palm leaf ash. *Structures*, **37**, pp. 1053–1068. https://linkinghub.elsevier.com/retrieve/pii/S2352012422000674. Accessed on: 20 Feb. 2022.
[19] Li, Q. et al., Effect of waste corn stalk ash on the early-age strength development of fly ash/cement composite. *Construction and Building Materials*, **303**, 124463, 2021.
[20] Akhayere, E. & Kavaz, D., Synthesis of silica nanoparticles from agricultural waste. *Agri-Waste and Microbes for Production of Sustainable Nanomaterials*, pp. 121–138, 2022.
[21] Cao, F., Qiao, H., Li, Y., Shu, X. & Cui, L., Effect of highland barley straw ash admixture on properties and microstructure of concrete. *Construction and Building Materials*, **315**, 125802, 2022.
[22] Odeyemi, S.O. et al., Mechanical properties and microstructure of high-performance concrete with bamboo leaf ash as additive. *Cleaner Engineering and Technology*, **6**, 100352, 2022.
[23] Bheel, N., Ali, M.O.A., Kirgiz, M.S., de Sousa Galdino, A.G. & Kumar, A., Fresh and mechanical properties of concrete made of binary substitution of millet husk ash and wheat straw ash for cement and fine aggregate. *Journal of Materials Research and Technology*, **13**, pp. 872–893, 2021.
[24] Hamid, Z. & Rafiq, S., An experimental study on behavior of wood ash in concrete as partial replacement of cement. *Materials Today: Proceedings*, **46**, pp. 3426–3429, 2021.
[25] Meko, B. & Ighalo, J.O., Utilization of Cordia Africana wood sawdust ash as partial cement replacement in C 25 concrete. *Cleaner Materials*, **1**, 100012, 2021.
[26] Raketh, M. et al., Sulfate removal using rubber wood ash to enhance biogas production from sulfate-rich wastewater generated from a concentrated latex factory. *Biochemical Engineering Journal*, **173**, 108084, 2021.
[27] Assiamah, S., Agyeman, S., Adinkrah-Appiah, K. & Danso, H., Utilization of sawdust ash as cement replacement for landcrete interlocking blocks production and mortarless construction. *Case Studies in Construction Materials*, **16**, e00945, 2022. https://linkinghub.elsevier.com/retrieve/pii/S2214509522000778. Accessed on: 21 Feb. 2022.
[28] Tijani, M.A., Ajagbe, W.O. & Agbede, O.A., Combined reusing of sorghum husk ash and recycled concrete aggregate for sustainable pervious concrete production. *Journal of Cleaner Production*, 131015, 2022. https://linkinghub.elsevier.com/retrieve/pii/S0959652622006503. Accessed on: 20 Feb. 2022.
[29] Koyuncu, D.D.E. & Okur, M., Investigation of dye removal ability and reusability of green and sustainable silica and carbon-silica hybrid aerogels prepared from paddy waste ash. *Colloids and Surfaces A: Physicochemical and Engineering Aspects*, **628**, 127370, 2021.
[30] Elango, K.S., Remya, P.R., Vivek, D., Gopi, R., Rajeshkumar, V. & Saravanakumar, R., Strength and durability studies on ficus exasperata leaf ash concrete. *Materials Today: Proceedings*, **37**(Part 2), pp. 999–1002, 2021.
[31] Al-Shwaiter, A., Awang, H. & Khalaf, M.A., Performance of sustainable lightweight foam concrete prepared using palm oil fuel ash as a sand replacement. *Construction and Building Materials*, **322**, 126482, 2022.

[32] Shahbazpanahi, S. & Faraj, R.H., Feasibility study on the use of shell sunflower ash and shell pumpkin ash as supplementary cementitious materials in concrete. *Journal of Building Engineering*, **30**, 101271, 2020.

[33] Oyebisi, S., Igba, T., Raheem, A. & Olutoge, F., Predicting the splitting tensile strength of concrete incorporating anacardium occidentale nut shell ash using reactivity index concepts and mix design proportions. *Case Studies in Construction Materials*, **13**, e00393, 2020.

[34] Kumar Yadav, A., Gaurav, K., Kishor, R. & Suman, S.K., Stabilization of alluvial soil for subgrade using rice husk ash, sugarcane bagasse ash and cow dung ash for rural roads. *International Journal of Pavement Research and Technology*, **10**(3), pp. 254–261, 2017. DOI: 10.1016/j.ijprt.2017.02.001.

[35] Ekasila, S., Vishali, B. & Anisuddin, S., An experimental investigation on compressive strength of concrete with rice husk and cow dunk ash using various curing methods. *Materials Today: Proceedings*, 2021.

[36] Abdul Razak, N.A., Othman, N.H., Mat Shayuti, M.S., Jumahat, A., Sapiai, N. & Lau, W.J., Agricultural and industrial waste-derived mesoporous silica nanoparticles: A review on chemical synthesis route. *Journal of Environmental Chemical Engineering*, **10**(2), 107322, 2022. https://linkinghub.elsevier.com/retrieve/pii/S2213343722001956. Accessed on: 23 Feb. 2022.

[37] Sarkar, J. et al., Synthesis of nanosilica from agricultural wastes and its multifaceted applications: A review. *Biocatalysis and Agricultural Biotechnology*, **37**, 102175, 2021.

[38] Shukla, S.S., Chava, R., Appari, S.A.B. & Kuncharam, B.V.R., Sustainable use of rice husk for the cleaner production of value-added products. *Journal of Environmental Chemical Engineering*, **10**(1), 106899, 2022.

[39] Goodman, B.A., Utilization of waste straw and husks from rice production: A review. *Journal of Bioresources and Bioproducts*, **5**(3), pp. 143–162, 2020.

[40] Mezan, S.O., al Absi, S.M., Jabbar, A.H., Roslan, M.S. & Agam, M.A., Synthesis and characterization of enhanced silica nanoparticle (SiO_2) prepared from rice husk ash immobilized of 3-(chloropropyl) triethoxysilanea. *Materials Today: Proceedings*, **42**, pp. 2464–2468, 2021.

[41] Imoisili, P.E., Ukoba, K.O. & Jen, T.C., Green technology extraction and characterisation of silica nanoparticles from palm kernel shell ash via sol-gel. *Journal of Materials Research and Technology*, **9**(1), pp. 307–313, 2020.

[42] Falk, G., Shinhe, G.P., Teixeira, L.B., Moraes, E.G. & de Oliveira, A.P.N., Synthesis of silica nanoparticles from sugarcane bagasse ash and nano-silicon via magnesiothermic reactions. *Ceramics International*, **45**(17), pp. 21618–21624, 2019.

[43] Khairul Hanif Mohd Nazri, M., & Sapawe, N., Effective performance of silica nanoparticles extracted from bamboo leaves ash for removal of phenol. *Materials Today: Proceedings*, **31**, pp. A27–A32, 2020.

[44] Mohamad, D.F. et al., Synthesis of mesoporous silica nanoparticle from banana peel ash for removal of phenol and methyl orange in aqueous solution. *Materials Today: Proceedings*, **19**, pp. 1119–1125, 2019.

[45] Sankar, S. et al., Biogenerated silica nanoparticles synthesized from sticky, red, and brown rice husk ashes by a chemical method. *Ceramics International*, **42**(4), pp. 4875–4885, 2016.

[46] Sankar, S., Kaur, N., Lee, S. & Kim, D.Y., Rapid sonochemical synthesis of spherical silica nanoparticles derived from brown rice husk. *Ceramics International*, **44**(7), pp. 8720–8724, 2018.

[47] Vaibhav, V., Vijayalakshmi, U. & Roopan, S.M., Agricultural waste as a source for the production of silica nanoparticles. *Spectrochimica Acta Part A: Molecular and Biomolecular Spectroscopy*, **139**, pp. 515–520, 2015.

[48] Mohd Nazri, M.K.H. & Sapawe, N., Removal of methyl orange over low-cost silica nanoparticles extracted from bamboo leaves ash. *Materials Today: Proceedings*, **31**, pp. A54–A57, 2020.

[49] Bokov, D. et al., Nanomaterial by sol-gel method: Synthesis and application. *Advances in Materials Science and Engineering*, ed. Z. Wang, 5102014, 2021. DOI: 10.1155/2021/5102014.

[50] Bahmani, S.H., Huat, B.B.K., Asadi, A. & Farzadnia, N., Stabilization of residual soil using SiO_2 nanoparticles and cement. *Construction and Building Materials*, **64**, pp. 350–359, 2014.

[51] Caputo, P., Porto, M., Angelico, R., Loise, V., Calandra, P. & Oliviero Rossi, C., Bitumen and asphalt concrete modified by nanometer-sized particles: Basic concepts, the state of the art and future perspectives of the nanoscale approach. *Advances in Colloid and Interface Science*, **285**, 102283, 2020.

[52] Karnati, S.R., Oldham, D., Fini, E.H. & Zhang, L., Surface functionalization of silica nanoparticles with swine manure-derived bio-binder to enhance bitumen performance in road pavement. *Construction and Building Materials*, **266**, 121000, 2021.

[53] Karnati, S.R., Oldham, D., Fini, E.H. & Zhang, L., Surface functionalization of silica nanoparticles to enhance aging resistance of asphalt binder. *Construction and Building Materials*, **211**, pp. 1065–1072, 2019.

[54] Karnati, S.R., Oldham, D., Fini, E.H. & Zhang, L., Application of surface-modified silica nanoparticles with dual silane coupling agents in bitumen for performance enhancement. *Construction and Building Materials*, **244**, 118324, 2020.

[55] Kamboozia, N., Saed, S.A. & Rad, S.M., Rheological behavior of asphalt binders and fatigue resistance of SMA mixtures modified with nano-silica containing RAP materials under the effect of mixture conditioning. *Construction and Building Materials*, **303**, 124433, 2021.

[56] He, H., Hu, J., Li, R., Shen, C., Pei, J. & Zhou, B., Study on rheological properties of silica nanofluids modified asphalt binder. *Construction and Building Materials*, **273**, 122046, 2021.

[57] Sukhija, M., Saboo, N., Yadav, A.K. & Rath, C., Laboratory study on the suitability of nano-silica as a modifier for asphalt binders. *Construction and Building Materials*, **302**, 124406, 2021.

[58] Crucho, J.M.L., das Neves, J.M.C., Capitão, S.D. & de Picado-Santos, L.G., Mechanical performance of asphalt concrete modified with nanoparticles: Nanosilica, zero-valent iron and nanoclay. *Construction and Building Materials*, **181**, pp. 309–318, 2018.

[59] Nazari, H., Naderi, K. & Moghadas Nejad, F., Improving aging resistance and fatigue performance of asphalt binders using inorganic nanoparticles. *Construction and Building Materials*, **170**, pp. 591–602, 2018.

[60] Long, Z., You, L., Tang, X., Ma, W., Ding, Y. & Xu, F., Analysis of interfacial adhesion properties of nano-silica modified asphalt mixtures using molecular dynamics simulation. *Construction and Building Materials*, **255**, 119354, 2020.

[61] Kie Badroodi, S., Reza Keymanesh, M. & Shafabakhsh, G., Experimental investigation of the fatigue phenomenon in nano silica-modified warm mix asphalt containing recycled asphalt considering self-healing behavior. *Construction and Building Materials*, **246**, 117558, 2020.
[62] Sezavar, R., Shafabakhsh, G. & Mirabdolazimi, S.M., New model of moisture susceptibility of nano silica-modified asphalt concrete using GMDH algorithm. *Construction and Building Materials*, **211**, pp. 528–538, 2019.
[63] Ghanoon, S.A. & Tanzadeh, J., Laboratory evaluation of nano-silica modification on rutting resistance of asphalt Binder. *Construction and Building Materials*, **223**, pp. 1074–1082, 2019.
[64] Gobinath, R., Raja, G., Prasath, E., Shyamala, G., Viloria, A. & Varela, N., Studies on strength characteristics of black cotton soil by using novel SiO_2 combination as a stabilizing agent. *Materials Today: Proceedings*, **27**, pp. 657–663, 2020.
[65] Hosseini, A., Haeri, S.M., Mahvelati, S. & Fathi, A., Feasibility of using electrokinetics and nanomaterials to stabilize and improve collapsible soils. *Journal of Rock Mechanics and Geotechnical Engineering*, **11**(5), pp. 1055–1065, 2019.

POSSIBILITY OF RECYCLING WIND TURBINE BLADES USING PLASMA TECHNOLOGY

VIKTORIJA GRIGAITIENĖ, ROLANDAS USCILA, VITAS VALINČIUS, ROMUALDAS KĖŽELIS, MINDAUGAS MILIEŠKA, DOVILĖ GIMŽAUSKAITĖ, VILMA SNAPKAUSKIENĖ & ŽYDRŪNAS KAVALIAUSKAS
Plasma Processing Laboratory, Lithuanian Energy Institute, Lithuania

ABSTRACT
The energy produced in wind farms is considered as mostly clean and environmentally friendly. However, a more detailed analysis of the process shows that this notion is not fully conclusive. Problems start already in the early stages of the manufacturing phase of wind turbines and become even more pronounced when they are disposed of at the end of their life-cycle. A review of the worldwide scientific and technical literature on the recycling of wind turbine blades has shown that there are currently no viable treatment methods for this waste. Global science and practice focus on traditional incineration and flame treatment technologies, with fragment mentions of alternative methods. However, traditional incineration and thermal treatments have serious disadvantages, making them almost impossible to obtain a final product suitable for further application. Most importantly, however, the thermal treatment process is not environmentally friendly due to the organic binders in the blades. Plasma treatment could be the alternative technology that can be used for the effective decomposition of all organic and inorganic materials and impurities in wind turbine blades that do not produce secondary waste. The treatment takes place without any special preparation of the raw material, except mechanical comminution. The energy released during the process can be used for the plant's own needs, as well as for municipal needs. Despite the complexity of the method, plasma facilities also have more advantages. In this work, the possibility of using plasma technology for the utilization of used wind turbine blades was investigated. Low-temperature plasma jet was used to recycle blade composites into gas and fiber melt.
Keywords: plasma processing, wind turbine blades, waste, composite materials, recycling.

1 INTRODUCTION

According to the Global Wind Energy Council (GWEC), the global installed wind power capacity rises every year, and it was 93.6 GW in 2022 [1]. The wind as an energy source helps to reduce greenhouse gas emissions, but because of the extreme loads, the average life of a wind turbine is about 20 years. The large number of wind turbines installed in Europe and other world countries at the end of the 20th century is near the end of their service and used wind turbine blades will have to be liquidated in the next few years: it will be the dismantling or modernization of the first wind farms. One of the challenges in wind energy management is to find a solution to recycle disposed wind turbine blades.

2 END-OF-LIFE MANAGEMENT OF WIND TURBINE BLADES

The research of University of Cambridge scientists estimates that there will be 43 million tons of retired wind turbine blades waste on our planet by 2050 [2]. At the present time, they're likely to pile up in landfills. WindEurope has called on the European Commission to ban the use of landfills for blades by 2025 (four European countries: Germany, Austria, The Netherlands and Finland, have banned it already) with a reason to help accelerate the development of sustainable recycling technologies.

A typical three-blade wind turbine consists of a tower, a nacelle, and three rotor blades. The tower provides structural support and is made of tubular steel and concrete. The nacelle contains steel and copper predominantly. Wind blades are attached to the rotor and nacelle

WIT Transactions on Ecology and the Environment, Vol 257,
www.witpress.com, ISSN 1743-3541 (on-line)
doi:10.2495/WMEI220061

and are made from composite materials. While up to 90% of wind turbine components can now be recycled and reused (i.e. concrete and metal parts), the problem remains the blades, which are made of non-metallic components that are not easily recyclable. Wind turbine blades are made of a composite of glass or carbon fiber and various polymers. This combination of fibers and polymers, also known as glass- or carbon-fibre-reinforced polymer composites, accounts for the majority of the blade material (60–70% fiber and 30–40% resin by weight) [3].

The development of the recycling method for commonly used blade materials started at global cement producer LafargeHolcim company (now Holcim Group) already in 2005. After shredding the blade, finely grained crushed blade dust material was added to the especially prepared fuel mix to bind it together. Keeping a kiln temperature around 900°C the resin burns, transforming the blade material into ashes, which was used as a filler in composite in the cement industry [4]. It helped significantly (by 27%) reduce the CO_2 output of the cement manufacturing process [5].

Nowadays, there are a few possible ways for dismantled wind turbine blades processing: landfill, incineration, or recycling. The leading recycling technologies which could be used in the waste management of wind turbine blades are mentioned in Fig. 1. When selecting landfill disposal, no material recovery is possible. The blades are sectioned into smaller pieces and placed in a landfill. This option is no way out. World countries aim strategy to reduce the quantity of waste going into landfills.

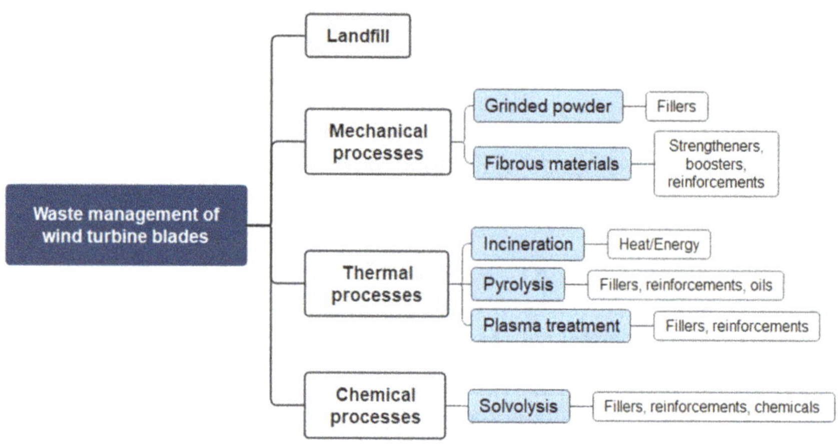

Figure 1: Recycling methods of wind turbine blades.

The development of alternative solutions and the creation of a recycling market are started already, either with an extended ban or with a sharp price increase for the landfill of the blades. The recycling technologies like mechanical shredding, chemical solvolysis (chemical reaction of solvent and solute that results in the formation of new compounds), and thermal decomposition are already being developed and improved.

When wind blades are recycled after the end of the work cycle by mechanical methods, they are only shredded and sorted. At the end of the process, the obtained milled mixture of composite pieces and powder is suitable for further application. However, due to their physical properties and chemical composition, the practical application of such wastes is limited [5], [6]. During mechanical blade treatment processes, a size reduction and separation into powder and fibrous fractions via cutting, crushing, shredding, grinding, or milling tools. Fine pieces are sieved out and used as fillers or reinforcements in building materials (cement, concrete) or as fuel for thermal waste processes. Theoretically, it has a wide range of potential applications for material recovery, but in addition, concrete made using shredded blades instead of crushed stones has unsatisfactory mechanical properties (lower stiffness and strength). The only solution is transforming fiberglass composites into small pellets, called EcoPoly (manufacturer Global Fibreglass Solutions). The pellets can be used as injectable plastics or highly waterproof panels that can be used in construction.

During the incineration (thermal process), the parts of the blades are incinerated at high temperatures – usually over 800°C. Organic substances are combusted and converted into ash and heat energy. This process is generally combined with energy production and heat recovery. The ash could be used as a recovering material as input for cement production or landfilled. The blades have high ash content, and burning composites create residues that can cause problems in the gas filtering systems. So, economically this treatment solution is not attractive.

During the pyrolysis (thermal process) method, the blades are cut into suitable pieces and decomposed using heating ovens in an inert atmosphere (450–700°C). Material is recovered in the form of fibers, which can be reused in other industries, such as glues, paints and concrete. Simultaneously, synthetic gas, which can later be combusted for electricity and heat recovery, and char products are gained during this process. The resin matrix can be transformed into pyrolysis oil. The disadvantage of this technology is that a significant amount of energy is needed to activate the pyrolysis process.

During solvolysis (chemical process), chemical solvents (usually water, alcohol, acid) are used to break the resin matrix bonds at elevated temperatures (300–650°C) and pressure conditions. After the treatment, fiber materials recovered have similar strength and could be reused in other applications, and the resins can be used for energy recovery. Regrettably, this technology focuses mainly on carbon fibers composites [7], [8].

All the technologies mentioned above allow to decompose wind turbine blades, but the final product is contaminated with impurities. Although these impurities could be easily removed, the possibilities for reusing the remaining fiber as an insulating material or producing other composites are limited. The main reason is that the glass crystallization process starts when exposed to temperatures higher than 500°C and fiberglass loses some of its physical and mechanical properties.

In this work, the possibility of using plasma technology to utilize used wind turbine blades was investigated. Low-temperature plasma jet was used to recycle blade composite into gas and fiber melt. This is a relatively expensive process, but it allows the complete decomposition of organic and inorganic substances present in the blades to a level that does not contain any toxic or other environmentally hazardous substances or compounds. A high density and non-equilibrium plasma jet was generated using 100 kW plasma torch. The glass melt formed during wind blade waste processing was leaching into new fibre employing a plasma-chemical reactor using the dynamic energy of the plasma jet. The whole process runs at atmospheric pressure and is friendly to the environment.

3 PLASMA WASTE TREATMENT TECHNOLOGY

The plasma waste treatment process can be a new alternative method for the recycling of wind turbine blades. It is related to intensive heat transfer between high-temperature plasma flow, dispersive heating particles and the environment. The experiments on fiber and granules production from different materials (sand, glass, zeolite, basalt, aluminum oxide, etc.) as well as the blades of dismantled wind turbine blades were performed by a direct current plasma-chemical reactor (Fig. 2) specially constructed for fiberization of raw ceramic and composite materials at Lithuanian Energy Institute [9].

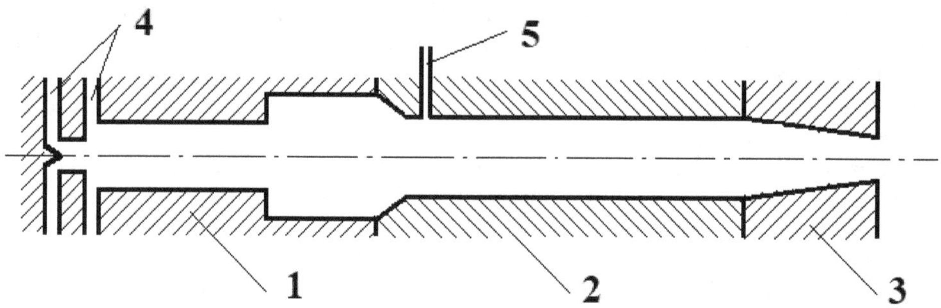

Figure 2: Experimental set-up of cylindrical plasma chemical reactor. 1 = Plasma torch; 2 = Reactor section; 3 = Outlet section; 4 = Working air feeding; 5 = Waste material feeding.

The electric arc ignited in the plasma torch heats up and ionizes the gas flow flowing through the reactor. When the powder of raw material is injected into the plasma flow, part of its energy is used to heat up the particles and melt them. The liquid melt mass flows along the plasma flow direction until it reaches the outlet of the reactor. As the final products, fibers and droplets are formed from the melt by the high-speed gas flow before they are cooled and solidified. The principal scheme of the waste recycling process by the plasma-chemical reactor is presented in Fig. 3.

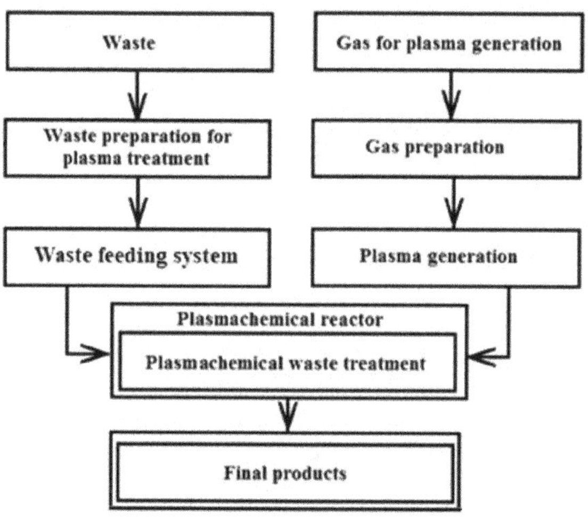

Figure 3: Scheme of blades waste treatment by the plasma torch.

Plasma torch, generating non-equilibrium plasma jet at atmospheric pressure, and supplying equipment necessary for high temperature jet generation are described in detail in

[9], [10]. The main operational parameters of the plasma torch used for experiments are: power supply (P) 70–120 kW, arc current (I) 150–300 A, arc voltage (U) 250–400 V, total airflow rate (G) 15–30 $g \cdot s^{-1}$, the average outlet temperature (T) 2500–3500 K, the average outlet velocity (v) 650–1250 $m \cdot s^{-1}$. The plasma chemical reactor was built of several separate sections 0.015 m in diameter from stainless steel with a length of 0.25 m (Fig. 2). The last section's exit diameter was 0.013 m. This allows to increase the outflow jet velocity and results in better melting process. All sections, as well as the plasma torch, were cooled with water. The wind turbine blades were shredded and milled down to 0.25 mm in size. Injected into the reactor, this powder was melted and fiberized in a reaction chamber connected to a plasma torch. The clogging at the outlet section during the recycling process was avoided regarding the high fluidity of the melt. Part of the melt was fiberized – pure and thin fiber was obtained (Fig. 4(a) and (b)). Part of the melt was also carried in the form of vitreous slag (Fig. 4(c) and (d)).

(a) (b)

(c) (d)

Figure 4: The morphology of fiber (a) and (b) and glassy slag (c) and (d), obtained from recycled wind turbine blades by plasma technology.

The plasma spray technology enables to join the processes of raw material melting, melt homogenization and formation of required material (fiber, nano-, micro-sized particles, or spherical ceramic granules) in a very short duration (0.1–0.5 ms) [10]. Produced fiber can be used as an insulation wool, a filter for ultrafine particles, or a concrete additive for strengthening building materials.

4 CONCLUSION

The global installed wind power capacity rises every year. The current and typical wind turbine blade life-cycle is about 20 years of service, and the waste stream after this time is dominated by the end-of-life blades, which becomes an urgent problem. The main ways of disposal are landfill, incineration or recycling. Currently, disposed wind turbine blades recycling and reuse technologies are limited, and most of them are still in the laboratory stage. A new plasma treatment method for recycling this type of waste is provided. Plasma spray technology demonstrates the ability to recycle wind turbine blades and gain products (fiber, vitreous slag) suitable for environmental, construction and insulating applications. The final products (fiber, vitreous slag) do not contain any additives or impurities. The process was realized by employing a specific plasma-chemical facility with a stream reactor and a linear DC plasma torch with up to 100 kW of power. The steady operating regime of the equipment is reached immediately after the start of the plasma torch. It was demonstrated that the final product's structure depends on the high viscosity of melted mass.

The results showed that plasma technology is an excellent tool for processing of wind turbine blades and allows to be optimistic for further tasks.

ACKNOWLEDGEMENT

This research has received funding from European Regional Development Fund (project number 01.2.2-LMT-K-718-01-0069) under a grant agreement with the Research Council of Lithuania (LMTLT).

REFERENCES

[1] The Global Wind Energy Council, Global Wind Report 2022. https://gwec.net/global-wind-report-2022. Accessed on: 12 Apr. 2022.

[2] Liu, P. & Barlow, C., Wind turbine blade waste in 2050. *Waste Management*, **62**, pp. 229–240, 2017.

[3] Mishnaevsky, L., Branner, K., Petersen, H.N., Beauson, J., McGugan, M. & Sørensen, B.F., Materials for wind turbine blades: An overview. *Materials*, **10**, p. 1285, 2017.

[4] Miceli, F., Circular economy: Use of wind turbine blades as combustible and mix material for cement production. Windfarmbop. https://www.windfarmbop.com. Accessed on: 12 Apr. 2022.

[5] Pickering, S.J., Recycling technologies for thermoset composite materials: Current status. *Composites Part A: Applied Science and Manufacturing*, **37**(8), pp. 1206–1215, 2006.

[6] Pławecka, K., Przybyła, J., Korniejenko, K., Lin, W.-T., Cheng, A. & Łach, M., Recycling of mechanically ground wind turbine blades as filler in geopolymer composite. *Materials*, **14**, p. 6539, 2021.

[7] Mattsson, C., André, A., Juntikka, M., Tränkle, T. & Sott, R., Chemical recycling of end-of-life wind turbine blades by solvolysis/HTL. *IOP Conference Series: Materials Science and Engineering. Proceedings of the 41st Riso International Symposium on Materials Science: Materials and Design for Next Generation Wind Turbine Blades*, Roskilde, Denmark, 7–10 Sep. 2020.

[8] Piñero-Hernanz, R., García-Serna, J., Dodds, C., Hyde, J., Poliakoff, M., Cocero, M.J., Kingman, S., Pickering, S. & Lester, E., Chemical recycling of carbon fibre composites using alcohols under subcritical and supercritical conditions. *J. Supercrit. Fluids*, **46**, pp. 83–92, 2008.

[9] Milieška, M., Kėželis, R., Valinčius, V., Grigaitienė, V. & Valatkevičius, P., Effect of outlet nozzle design on the morphology and microstructure of micro-and nanostructure fibre formed by means of plasma technology. *Radiation Interaction with Material and its Use in Technologies: 3rd International Conference*, Kaunas, Lithuania, 20–23 Sep. 2010. Kaunas: Technologija, pp. 371–374, 2010.

[10] Valinčius, V., Snapkauskiene, V., Kezelis, R., Valinciute, V. & Mecius, V., Preparation of insulating refractory materials by plasma spray technology. *J. of High. Temp. Mat. Proc.*,**10**, pp. 365–375, 2006.

RECYCLING OF LOW-DENSITY POLYETHYLENE WASTE TO PRODUCE ECO-FRIENDLY GYPSUM COMPOSITES

MARÍA I. ROMERO-GÓMEZ[1], PALOMA RUBIO-DE-HITA[1], MANUEL A. PEDREÑO-ROJAS[2], MARÍA J. MORALES-CONDE[1] & FILOMENA PÉREZ-GÁLVEZ[1]
[1]Departamento de Construcciones Arquitectónicas I, Escuela Técnica Superior de Arquitectura, Universidad de Sevilla, Spain
[2]Departamento de Urbanística y Ordenación del Territorio, Escuela Técnica Superior de Arquitectura, Universidad de Sevilla, Sevilla, Spain

ABSTRACT

Over 367 million metric tons of plastic were produced worldwide in 2020. Within the most consumed standard plastics, low-density polyethylene (LDPE) is the second resin most demanded in Europe. Far from reducing its consumption, this continues to grow despite environmental impact. Due to the low recycling rate of LDPE waste (~35%), coming mainly from post-consumer packaging and lightweight bags, most parts of this waste end up in the environment. As an alternative to restrained governmental steps, i.e. extra taxes to reduce single-use plastic consumption, new solutions for the recycling and reuse of plastic waste are being developing from different economic sectors. Because of the worldwide significance of the construction sector, which has a large consumption of natural resources and production of waste, this sector plays a critical role within the new models of circular economy. In light of this, and committed to the sustainable development goals adopted by the United Nations, the present work intends to determine the feasibility of using LDPE waste sourced from plastic bags to produce eco-friendly gypsum-based composites. Four replacement levels of plastic waste were chosen for analysis – 0.25%, 0.5%, 1%, 5% by weight of gypsum. The experimental campaign consisted of analysing the resulting compounds based on density, mechanical behaviour (flexural and compressive strength), as well as evaluating their water absorption by capillary action and adsorption by exposing them to continuous moisture. Comparing this to the control material, lighter composites with a decrease of mechanical strength were obtained. However, all the values of flexural and compressive strength exceeded the minimums established by standard. Also, the LDPE-containing composites presented lower water absorption and adsorption capacity (up to ~50% when compared to the control material). So it could be inferred that the waste material can be considered as a viable partial substitute to the current commercial gypsum for the manufacturing of gypsum products, i.e. plasterboards.
Keywords: LDPE waste, gypsum plaster, mechanical properties, water-resistance properties, eco-friendly material.

1 INTRODUCTION

Plastic waste management stands as one of the main worldwide concerns to minimise the impact of human activities on the global environment. Over 55 million metric tons of plastic were produced in Europe in 2020. However, the amount of plastic post-consumer waste collected was lower than 50% of the total produced, from which 23% was still sent to landfill [1]. Within the most consumed standard plastics, low-density polyethylene (LDPE) is the second resin more demanded in Europe [1]. The lightness, durability, low thermal conductivity, impermeability and low cost are some of the most notable properties that make that far from reducing the consumption of LDPE, this continues to grow despite environmental impact [2].

Due to the low recycling rate of LDPE waste (~35%), coming mainly from post-consumer packaging and lightweight bags (single-use products), most part of this waste end up in the environment. Recently research showed that plastic litter exposure to external mechanical forces, as well as solar radiation may result in the transformation of this waste to microplastics [3]. Thus, because of the high resistance of LDPE to natural degradation,

WIT Transactions on Ecology and the Environment, Vol 257, © 2022 WIT Press
www.witpress.com, ISSN 1743-3541 (on-line)
doi:10.2495/WMEI220071

considerable damage toward the terrestrial and marine ecosystems is caused [4], [5]. In an attempt to look for effective solutions to reduce environmental pollution derived from single-use plastics, restrained governmental steps, i.e. extra taxes added to the consumption of this type of products, have been stablished [6]. Nevertheless, there is a need to find suitable ways to reuse or recycle plastic, primarily those with a low recycling rate in comparison with their production (i.e. low density polyethylene), from different economic sectors. Because of the worldwide significance of construction sector, which means, great consume of natural resources and production of waste, this sector plays a critical role within the new models of circular economy.

In this line, numerous research on the recycling of plastic waste to produce new construction materials for a variety of sustainable building applications, have been conducted by architectural and civil engineering sectors.

Related to the use of recycled LDPE waste to develop new building materials, interesting studies were published in concrete manufacture field. The influence of shredded LDPE from plastic bags used as partial substitution of fine aggregate on the behaviour of concrete was analysed by Jain et al. [7]. As a result, the workability, density, mechanical strength and modulus of elasticity of concrete samples decreased with increase in the plastic waste content, while penetrability to water rose. Also, Poonyakan et al. [8] found a significant reduction of thermal conductivity of LDPE-containing concrete composites at 30% volume fraction of LDPE content, when compared with the plain concrete.

Similarly, the influence of adding fine LDPE waste in cement composites, as partial replacement of sand, on density, mechanical strength and durability properties of plastic cement mortars was evaluated by several studies [9], [10]. Based on the prediction model for cement composites made with LDPE waste designed by Ohemeng and Ekolu [11], the LDPE-containing mortars' flexural and compressive strength were adversely affected with the increase of percentage of plastic waste. However, the standard strength requirements were satisfied by composites with LDPE replacement levels of up to 60% by volume. Furthermore, the possible application of this new cement-based materials with LDPE addition to manufacture paver bricks was researched by Reddy et al. [12].

On the other hand, encouraging alternative eco-friendly solutions to commercial gypsum conglomerate by proposing new recycling ways for diverse type of plastic residues without efficient management at present, have been published. Lightweight materials with enhanced thermal properties made with gypsum and plastics residues from CDW (EPS and XPS, respectively) were developed by San-Antonio-González et al. [13], [14]. In this line, Pedreño-Rojas et al. [15] verified the feasibility of using PC waste to produce lighter and mechanically improved gypsum composites, by using up to a 60wt% PC waste addition. Moreover, the improved water-resistant properties of gypsum composites with the incorporation of plastic waste from rejected cables was analysed by Vidales-Barriguete et al. [16]. Other interesting works sought to find a solution to LDPE plastic pollution and reduce the amount of raw material needed to make enhanced mechanical and thermal LDPE-laterite-gypsum bricks by using reprocessed LDPE, calcium phosphate (CaP), phosphogypsum residues and biochar was presented by Azeko et al. [17].

In spite of several studies have evaluated the influence of LDPE waste addition on physico–mechanical and thermal behaviour of concretes and cement-based composites, few research have been developed on the use of LDPE waste as partial replacement to gypsum binder matrixes Therefore, committed to the Sustainable Development Goals adopted by the United Nations, the present work intend to determine the feasibility of using LDPE waste sourced from plastic bags to produce eco-friendly gypsum-based composites. The experimental campaign planed in this work consisted of analysing the resulting compounds based on density, mechanical behaviour (flexural and compressive strength), as well as, evaluating their water absorption by capillary action and adsorption by exposing them to continuous moisture.

2 MATERIALS AND METHODS

2.1 Materials

The gypsum composites developed in this work were prepared by using the following materials:

- Controlled setting gypsum for construction (B1), according to the standard EN 13279-1 [18].
- Regular tap water in accordance with Council Directive 98/83/EC.
- Low density polyethylene (LDPE) waste aggregate (Fig. 1(a)), obtained from discarded lightweight plastic bags. This residue was collected, processed, and provided by "Reciclados La Red" recycled plant from Seville (Spain). The recycled plastics bags were crushed by using the recycling company's industrial equipment with an output of particles measuring up to ~20 mm (particle size distribution of the ground LDPE waste in Fig. 1(b)).

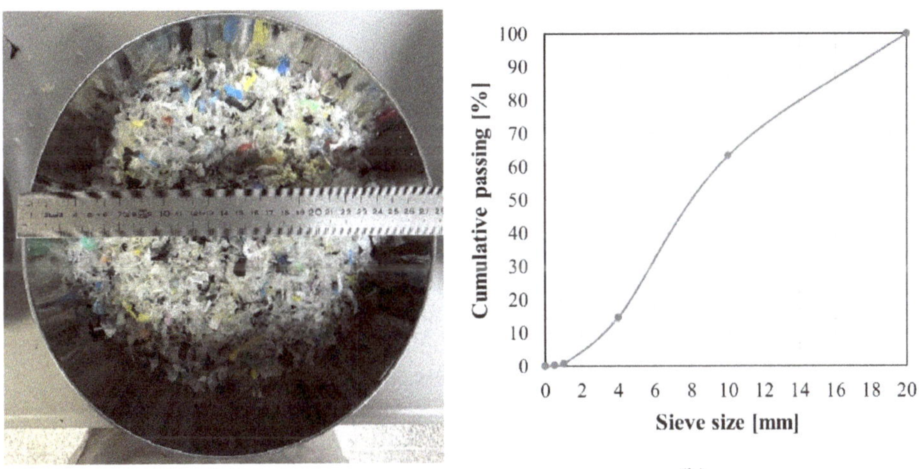

(a) (b)

Figure 1: LDPE waste used. (a) LDPE waste particles ready to be mixed; and (b) Particle size distribution.

2.2 Samples preparation

The guidelines set by standard EN 13279-2 [19] were followed to prepare the samples evaluated at this work. Four plastic waste replacement levels were chosen: 0.25%, 0.5%, 1% and 5% by weight of gypsum (wt%). The maximum percentage of LDPE waste was conditioned by the water/gypsum ratio (W/G) of 0.55 fixed for all the mixtures to guarantee a correct workability of the mixtures without using additives.

A manually procedure was used to prepare the mixes, according to the proportions of material summarized in Table 1. In the first place, LDPE waste aggregate was dry mixed with gypsum powder to avoid a heterogeneous dispersion of the particles through the gypsum matrix. Subsequently, water was gradually added to the rest of blend components and mixed for 2 minutes until a homogeneous state.

Table 1: Mix proportions for each group of six specimens (40 × 40 × 160 mm^3).

Mix code	Gypsum (g)	Water (g)	W/G ratio	LDPE waste (g)
G/CM	1,800	990	0.55	–
G/LDPE/0.25	1,796	988	0.55	4.5
G/LDPE/0.5	1,791	985	0.55	9
G/LDPE/1	1,782	980	0.55	18
G/LDPE/5	1,710	941	0.55	90

After completing this procedure, the mixtures were poured into prismatic moulds of 40 × 40 × 160 mm^3 to produce six specimens by serie. A total of 24 samples were prepared, together with the control specimen (G/CM) without plastic waste addition.

2.3 Test methods

Once samples were removed from the moulds, these were cured according to conditions stablished by standard EN 13279-2 [19]. Thus, specimens were stored for seven days in a dry chamber at a temperature of 24°C and a relative humidity of 50 ± 1%. Then, they were placed in an oven at 40 ± 2°C to constant mass and cooled to laboratory temperature in a desiccator. Concluded the curing period, the gypsum composites were subjected to the test methods described below:

- *Dry bulk density*: calculated after measuring the dry weight and volume of the test samples, as per standard EN 13279-2 [19].
- *Flexural strength:* based on the procedure explained in EN 13279-2 [19], six samples for mixture were subjected to three-point bending test by applying a progressive centred load with a speed of 10 N/s, until reach the breaking point.
- *Compressive strength*: the 12 broken test sections obtained from the previous flexural strength test were used to carry out the compressive strength test, as per standard EN 13279-2 [19]. A load was applied in the central plane of the sample with a speed of 20N/s, until specimen failure.
  The automatic multi-test press MCO-30 equipment, with a 300 KN load capacity was used to develop both test methods.
- *Water absorption by capillary action:* as determined by standard RILEM TC 25-PEM [20], three samples per serie were placed in a vertical position inside a container, with a distance 10 ± 1 mm of water from the bottom. During a testing period of 10 minutes, the water level in the specimens' surface was measured every minute.
- *Hygroscopicity range:* to developed this non-standardized test, designed by Del Río-Merino [21] in her thesis, samples were undergone to constant moisture, by placing them in a wet chamber for five days at a temperature of 21 ± 2°C and a RH of 72 ± 5%. Afterwards, the difference of samples weight gain was measured.

3 RESULTS AND DISCUSSIONS

The resulting data obtained for the experimental campaign to previously described to characterize the new LDPE-containing composites are summed up in Table 2.

Table 2: Summary table with gypsum composites' density and mechanical strength tests results.

Specimens	Density(kg/m^3) (CoV (%))	Flexural strength (MPa) (CoV (%))	Compressive strength (Mpa) (CoV (%))
G/CM	1,205 (1.44)	3.39 (7.59)	7.71 (5.61)
G/LDPE/0.25	1,195 (0.66)	2.35 (7.46)	5.50 (2.13)
G/LDPE/0.5	1,190 (0.65)	2.30 (5.17)	4.90 (5.00)
G/LDPE/1	1,185 (0.57)	2.17 (5.75)	4.59 (3.89)
G/LDPE/5	1,183 (1.12)	2.03 (10.05)	3.02 (5.71)

3.1 Dry bulk density

Regarding to the results observed in the Fig. 2, density values of gypsum-based composited, in all cases, were reduced as the replacement of LDPE waste increased. Compared to reference sample, the most pronounced fall of density was produced when a percentage of 5wt% LDPE waste content (~2% decrease) was added to the mix. However, it should be noted that the highest variation of density values between the control material and the LDPE-containing samples was corresponded to G/LDPE/1 compound. The partial replacement of gypsum binder matrix by LDPE waste with lower density (2650 kg/m^3 vs. 920 kg/m^3, respectively) could explained the declining trend observed related to dry bulk density.

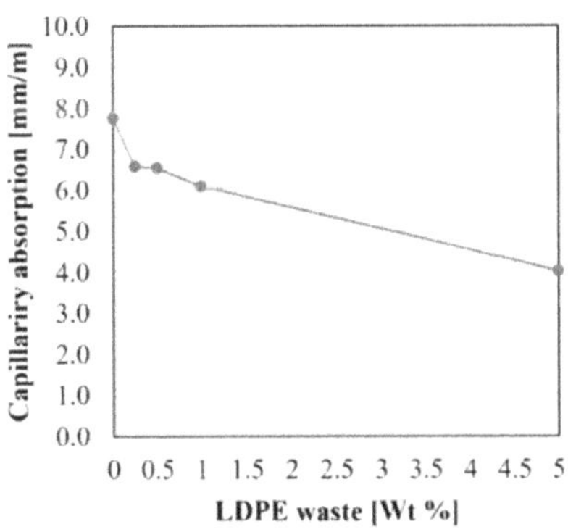

Figure 2: Dry bulk density values (kg/m^3) of gypsum/LDPE composites.

3.2 Mechanical properties

Expected results were obtained for the flexural and compressive strength tests, as it is shown in Fig. 3. In both cases, the mechanical properties of the new gypsum composites were decreased when the percentage of LDPE addition rose. Concerning to flexural strength, a reduction up to ~40% was reached by composites with a 5wt% LDPE content. At the same way, the higher the percentage of plastic waste aggregate was added to the blends, the lower the compressive strength values were achieved in relation to the reference material. The maximum fall of ~60% was reached by G/LDPE/5 composites. The behaviour presented by new LDPE-containing gypsum composites pointed out that the cohesion at the interfacial transition zone (ITZ) between the LDPE waste particles and the gypsum matrix it is not as strong as that in between the products of hydration and is likely to be a point of failure. Nevertheless, in both flexural and compressive strength tests, all the values were within the minimums 1 MPa and 2 MPa, respectively, stablished by standard EN 13279-2 [19].

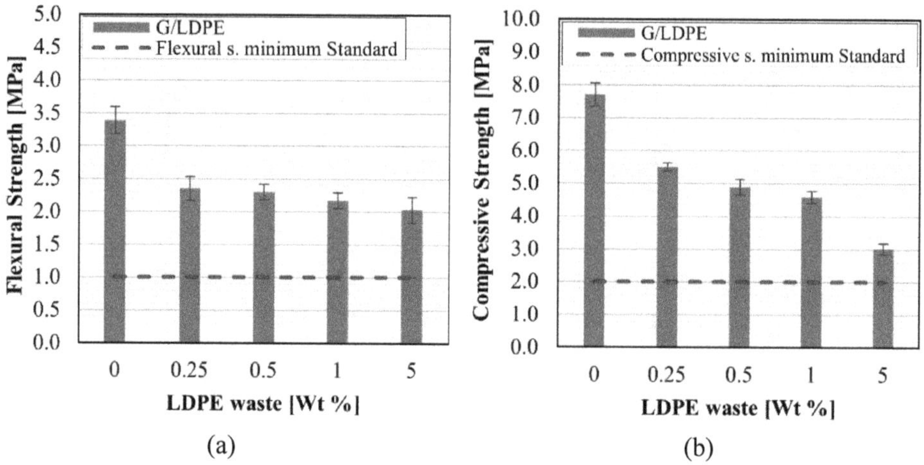

Figure 3: Flexural and compressive strength values (MPa) of gypsum/LDPE composites.

3.3 Water absorption by capillary action

Table 3 showed the data obtained from the water absorption by capillary action test. It can be observed a widespread significant water uptake reduction for the new LDPE-containing composites. The highest decline (up to ~48%) was reached by composites with 5wt% LDPE addition, when compared to the reference gypsum. Concerning to the average water absorption by capillary action (Fig. 4), the lowest value was that of G/LDPE/5, (i.e. 4 mm/min). So, it could be inferred that the increasing presence of LDPE particles imparts an impervious property thereby slowing down the advance of water.

Table 3: Capillary water absorption test results, measured for 10 min.

	Capillary absorption by minute (mm)										Total capillary absorption (mm)	Δ Capillary absorption compared to ref. (%)
Specimens	1	2	3	4	5	6	7	8	9	10		
G/CM	39	6	7	5	6	5	3	7	2	2	85.3	–
G/LDPE/0.25	28	9	6	6	5	5	4	4	3	3	72.3	–15.23
G/LDPE/0.5	31	8	6	5	6	4	2	4	4	2	72.0	–15.63
G/LDPE/1	25	8	6	5	6	3	5	4	3	3	67.0	–21.48
G/LDPE/5	20	3	4	3	3	3	2	2	2	2	44.3	–48.05

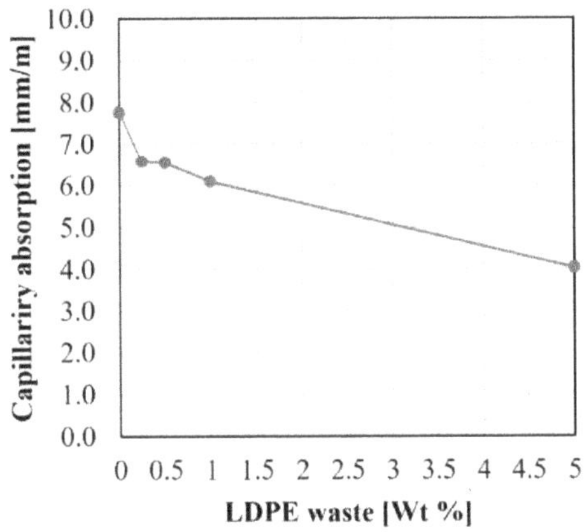

Figure 4: Average water uptake by capillarity (mm/m) of gypsum/LDPE composites.

3.4 Hygroscopicity range

After exposing the gypsum composites developed to continuous moderate humidity until 72% RH the results are summarised in Table 4. As it can be observed, the higher the LDPE replacement level was, the lower water adsorption capacity was. Therefore, G/LDPE/5 composite achieved the highest reduction on water adsorption capacity, about 91% lower than the control sample. This trend is similar to that observed in Section 3.3. So, in can be concluded that the growing addition of an impervious material (LDPE waste) lead to an increase of composites' resistance to water movement.

Table 4: Hygroscopicity of the gypsum composites exposed to continuous moderate humidity until 72% RH.

Specimens	Weight before the test: Dry (g)	Weight after the test: 72% RH (g)	Water adsorption (%)
G/CM	307.61	314.18	+2.14
G/LDPE/0.25	304.72	319.27	+1.99
G/LDPE/0.5	305.87	314.35	+1.78
G/LDPE/1	303.41	303.98	+1.07
G/LDPE/5	302.92	303.46	+0.18

4 CONCLUSIONS

From above research it can be inferred the feasibility of recycling LDPE waste obtained from lightweight plastic bags as aggregate to partially replace the raw material (i.e. gypsum) in the production of eco-friendly gypsum composites. By considering the results obtained from the characterization tests developed, the following conclusions were drawn:

- No alterations on gypsum mixtures workability were observed after adding of up to 5wt% LDPE waste, which allowed to maintain a 0.55 w/g ratio without adding water reducing admixtures.

- Slight reduction density in dry state of the LDPE-containing gypsums was observed, when compared with the control material (~2% decrease with 5% LDPE waste content).
- Mechanical strength properties showed a widespread decreased with the increasing incorporation of LDPE waste to the gypsum composites. Compared to the control material, the new gypsum-based composites presented a progressive decline trend in flexural and compressive, which is greater in mixtures with a higher level of LDPE waste content (up to ~40% and ~60% decrease, respectively, corresponding to G/LPDE/5 compounds). Nevertheless, all the values were above the minimums indicated in the standard.
- Water absorption and adsorption capacity were reduced after adding impervious LDPE waste to gypsum mixes. This decline observed in both properties, was more noticeable with increasing LDPE waste content (~48% and ~91% reduction, receptively when a proportion of 5wt% LDPE waste was used, compared to control material).

Finally, it can be concluded the viability of producing eco-friendly substitute materials for commercial gypsum used for construction applications included in circular economy strategies (i.e. gypsum drywall), that comply with all the mechanical regulatory requirements and, with improved water-resistance performance, when compared to the plain gypsum without plastic. Likewise, an alternative proposal of waste valorisation to promote the efficient disposition of LDPE waste to facilitate their recycling is presented.

ACKNOWLEDGEMENTS

The author, M.I. Romero-Gómez, wishes to acknowledge the financial support provided by the FPU Program of Spanish Ministry of Science, Research and Universities (FPU 18/02405).

REFERENCES

[1] Plastics Europe, Plastics the facts 2021. An analysis of European plastics production, demand and waste data. https://plasticseurope.org/knowledge-hub/plastics-the-facts-2021/. Accessed on: 10 May 2022.

[2] Wypych, G. & Wypych, G., LDPE low density polyethylene. *Handbook of Polymers*, pp. 178–184, 2016.

[3] Steensgaard, I.M., Syberg, K., Rist, S., Hartmann, N.B., Boldrin, A. & Foss, S., From macro- to microplastics: Analysis of EU regulation along the life cycle of plastic bags. *Environ. Pollut.*, **224**, pp. 289–299, 2017.

[4] Hadiuzzaman, M., Salehi, M. & Fujiwara, T., Plastic litter fate and contaminant transport within the urban environment, photodegradation, fragmentation, and heavy metal uptake from storm runoff. *Environ. Res.*, **212**, 113183, 2022.

[5] Sarker, R.K., Chakraborty, P., Paul, P., Chatterjee, A. & Tribedi, P., Degradation of low-density polyethylene (LDPE) by Enterobacter cloacae AKS7: A potential step towards sustainable environmental remediation. *Arch. Microbiol.,* **202**(8), pp. 2117–2125, 2020.

[6] European Union, Directive (EU) 2019/904 of the European Parliament and of the Council of 5 June 2019 on the reduction of the impact of certain plastic products on the environment. (Text with EEA relevance.)

[7] Jain, A., Siddique, S., Gupta, T., Sharma, R.K. & Chaudhary, S., Utilization of shredded waste plastic bags to improve impact and abrasion resistance of concrete. *Environ. Dev. Sustain.*, pp. 1–26, 2018.

[8] Poonyakan, A., Rachakornkij, M., Wecharatana, M. & Smittakorn, W., Potential use of plastic wastes for low thermal conductivity concrete. *Materials (Basel)*, **11**(10), 2018.

[9] Ghernouti, Y. & Rabehi, B., Strength and durability of mortar made with plastics bag waste (MPBW). *Int. J. Concr. Struct. Mater.*, **6**(3), pp. 145–153, 2012.

[10] Jassim, A.K., Recycling of polyethylene waste to produce plastic cement. *Procedia Manuf.*, **8**, pp. 635–642, 2017.

[11] Ohemeng, E.A. & Ekolu, S.O., Strength prediction model for cement mortar made with waste LDPE plastic as fine aggregate. *Journal of Sustainable Cement-Based Materials*, **8**(4), pp. 228–243, 2019.

[12] Reddy, C.S.K., Chandra Kumar, B.S. & Asadi, S.S., Utilization of low density polyethylene waste in the manufacturing of paver brick. *Int. J. Recent Technol. Eng.*, 7(6C2), pp. 62–67, 2019.

[13] San-Antonio-González, A., Del Río-Merino, M., Viñas-Arrebola, C. & Villoria-Sáez, P., Lightweight material made with gypsum and extruded polystyrene waste with enhanced thermal behaviour. *Constr. Build. Mater.*, **93**, pp. 57–63, 2015.

[14] San-Antonio-González, A., Del Río-Merino, M., Viñas-Arrebola, C. & Villoria-Sáez, P., Lightweight material made with gypsum and EPS waste with enhanced mechanical strength. *J. Mater. Civ. Eng.*, **28**(2), 04015101, 2015.

[15] Pedreño-Rojas, M.A., Morales-Conde, M.J., Pérez-Gálvez, F. & Rubio-de-Hita, P., Influence of polycarbonate waste on gypsum composites: Mechanical and environmental study. *J. Clean. Prod.*, **218**, pp. 21–37, 2019.

[16] Vidales-Barriguete, A., Atanes-Sánchez, E., Del Río-Merino, M. & Piña-Ramírez, C., Analysis of the improved water-resistant properties of plaster compounds with the addition of plastic waste. *Constr. Build. Mater.*, **230**, 116956, 2020.

[17] Azeko, S.T., Arthur, E.K., Minh, D.P., Lyczko, N., Nzihou, A. & Soboyejo, W.O., Mechanical and thermal properties of sustainable composite building materials produced by the reprocessing of low-density polyethylene, biochar, calcium phosphate, and phosphogypsum wastes. *J. Mater. Civ. Eng.*, **34**(2), 04021457, 2021.

[18] EN 13279-1, Gypsum binders and gypsum plasters – Part 1: Definitions and requirements, 2006.

[19] UNE-EN 13279-2, Gypsum binders and gypsum plasters – Part 2: Test methods, 2006.

[20] RILEM TC 25-PEM, Recommended tests to measure the deterioration of stone and to assess the effectiveness of treatment methods. *Mater. Struct.*, **13**(75), pp. 175–253, 1980.

[21] Del Río-Merino, M., Elaboración y aplicaciones constructivas de paneles prefabricados de escayola aligerada y reforzada con fibras de vidrio E y otros aditivos. Doctoral dissertation, Architecture, E.T.S. de Edificación, UPM, 1999.

SECTION 3
STRATEGIES FOR WASTE MANAGEMENT: EU AND NON-EU CASE STUDIES

COVID-19 AND MUNICIPAL SOLID WASTE MANAGEMENT

ELENA CRISTINA RADA
Department of Theoretical and Applied Sciences, University of Insubria, Italy

ABSTRACT
The present work analyses the scientific literature available on the impact of the SARS-CoV-2 in the sector of municipal solid waste. After a bibliometric analysis based on data taken from Scopus®, the analysis deals with effects on waste composition, waste quantity, collection, and treatment. As expected, results show that the most productive authors on this topic belong to scientific bodies located in the countries more affected from COVID-19 in the world. Moreover, different strategies of international journals resulted in an unbalanced concentration of papers on this topic. Effects have been observed concerning municipal solid waste composition and amount (mostly for the role of masks and packaging). Impacts on management and circular economy are discussed too.
Keywords: COVID-19, circular economy, municipal solid waste, SARS-CoV-2, urban waste, waste management.

1 INTRODUCTION

During the first period of the COVID-19 pandemic, one of the criticalities to be faced with and to be solved quickly was the defence of the correct collection and treatment/disposal of municipal solid waste (MSW) to avoid risky side effects from unhealthy practices. A quick response went from the bodies responsible of health protection at national level through the issue of guidelines. In the same time an international activity evolved at research level. The present article zooms on this second aspect in order to understand which contribution came to fight SARS-CoV-2 in a short amount of time.

2 MATERIAL AND METHODS

The approach used in the present paper is based on four steps aimed to valorise the contents of the international scientific databases that significantly evolved in the latest decades. The steps are described as follows:

- Select an international scientific database suitable for a bibliometric analysis on relationships between COVID-19 and MSW management;
- Analyse the bibliometric data on the above-mentioned topic;
- Analyse the contents to find relevant issues to understand the impact of the pandemic on the sector of MSW, including sustainability and circular economy;
- Check if a qualified database can cover all the aspects of interest for MSW and pandemic.

3 RESULTS AND DISCUSSION

The chosen databased was Scopus® because of its wide coverage of peer review international journals and conferences. Indeed, the aim was to analyse the scientific response of researchers to the pandemic issues, specifically in the sector of municipal solid waste. This database was accessed in May 2022 to extract data referred to two queries: COVID-19 and “municipal solid waste” and COVID-19 and “urban waste”, and the results are reported in Fig. 1.

WIT Transactions on Ecology and the Environment, Vol 257, © 2022 WIT Press
www.witpress.com, ISSN 1743-3541 (on-line)
doi:10.2495/WMEI220081

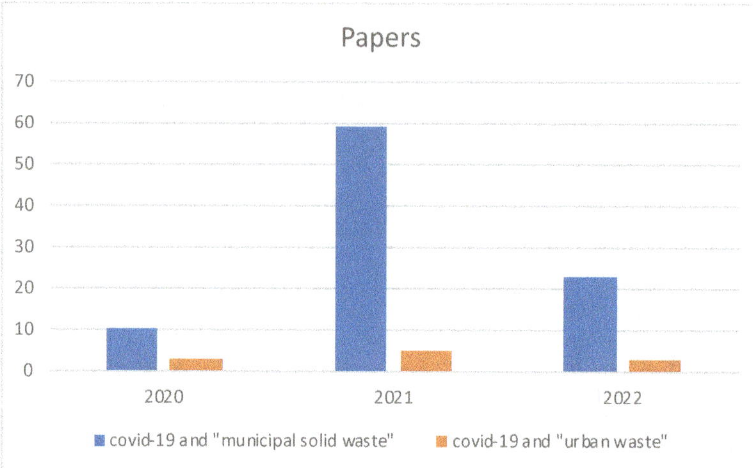

Figure 1: Papers published on the two queries in Scopus®.

The results from the Scopus® databased showed 92 documents connected with the first query and 11 connected with the second one [1]–[103]. The expectation is towards a slight increase from 2021 to 2022 possibly as the main questions to manage MSW during the pandemic has found an answer. The papers on the second query were published in 11 different journals [93]–[103]. Some of the journals that published papers on the first query have more than one paper. Indeed, *Science of the Total Environment* dedicated a special issue on SARS-CoV-2 and collected 15 papers [1], [4], [5], [10], [16], [24], [25], [31]–[33], [36], [43], [63], [83], [84]. The other two journals that attracted papers connected with the first query are *Resources, Conservation and Recycling* [81], [82], [85], [86], [102] and *Environmental Science and Pollution Research* [14], [21], [27], [44], [48], each of them having five papers published. Four other journals, *Waste Management and Research* [8], [51], [57], *Journal of Hazardous Materials* [9], [11], [12], *Environmental Research* [23], [61], [89] and *Journal of Environmental Management* [53], [63], [78] published three papers each. Regarding the proceedings from different conferences, in the selected period only six papers were published [37], [64]–[66], [71], [92].

As an overview, it can be noticed that:

- 103 documents since June 2020, showing the typical delay from submission and acceptance of a papers and from these only five also considered the circular economy concepts [3], [25], [65], [72], [87];
- 79 were open access (a big part of the ones published on the first query and all published on the second one) showing the attention to guarantee a quick dissemination of the results of research around the world in order to help decision makers during the pandemic.

In order to understand the relevance of the scientific production, it must be underlined that in 2021 the papers in Scopus® on MSW or urban waste were 2,389. In practice the topic is belonging to a niche. May be the explanation can be done looking at Fig. 2(a), showing the affiliations of the corresponding authors of the articles in Scopus®: the presence of China (eight documents) and Italy (three documents) is visible as expected. In practice, the main production comes mostly from two countries that had to face the hardest impact of COVID-

19. Again, it is not surprising that an Italian author is present in Fig. 2(b) reporting the researchers with the highest production on the topic [65], [83], [91].

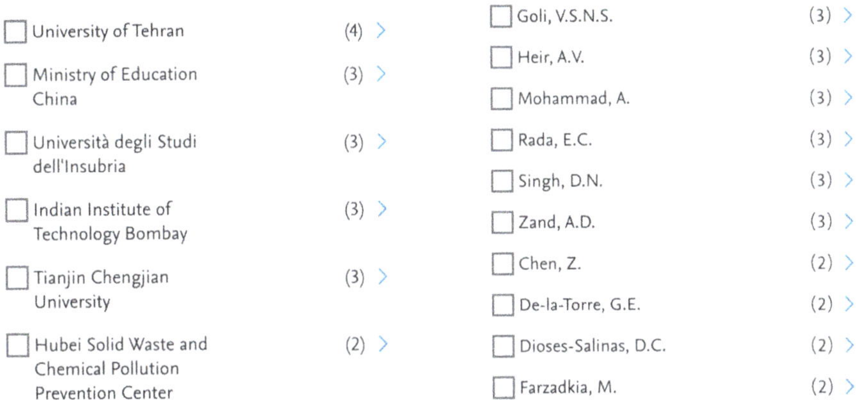

(a) (b)

Figure 2: (a) Affiliations of the corresponding authors who published on the topic in Scopus®; and (b) Main authors who published articles in Scopus® on the topic.

From another point of view, Table 1 is useful to see that articles on this topic were accepted in Scopus® indexed journals of high impact factor (Web of Science IF).

Table 1: Journals with the highest presence of papers on the topic and latest impact factor.

Science of the Total Environment	IF = 7.963 (WoS)
Environmental Science and Pollution Research	IF = 4.223 (WoS)
Resources, Conservation and Recycling	IF = 9.930 (WoS)
Waste Management and Research	IF = 3.549 (WoS)
Journal of Hazardous Materials	IF = 10.390 (WoS)
Environmental Research	IF = 6.498 (WoS)
Journal of Environmental Management	IF = 6.789 (WoS)

Zooming in on the contents of the articles, the target is different from the guidelines [104] (and their updates) that characterised the first period of pandemic: in practice the European Union issued quickly strategic guidelines in order to prevent a crisis in a sector (MSW collection and treatment) that can turn to be critical if the exposure of the operators went out of control. Looking to that period, the guidelines acted successfully.

The Scopus® literature followed a parallel pathway with time depending on the peer review lasting of the related journals. The uncertainty of the per review lasting pushed the authors to introduce in the text the concept that results will be useful for future pandemics. The presence of articles oriented to low–medium-income contexts is significant. Indeed, differently from EU and EU-like countries the MSW sector is far from to be optimised: EU guidelines could not be applied.

In Table 2 a resume of the most significant contents from the 103 documents extracted from Scopus® is presented.

Table 2: Indications for/from non-EU and EU like countries.

Indications referred to the pandemic period	Ref.
Attention must be exploited to set optimised MSW guidelines for the future	[15]
SARS-CoV-2 RNA detection in landfill leachate helps as surveillance strategy	[4]
Food delivery is contributing to a significant increase of plastic waste during the pandemic	[93]
Education and training were a significant predictor of health-protective behaviour	[22]
The presence of SARS-CoV-2 and other pathogenic microorganisms in sewage sludge, wastewater, and landfill leachate can hamper the possibility to ensure safe water and public health in economically marginalized countries	[23]
Compared to 2019, prolonged lockdowns caused larger decreases in the quantity of commercial and construction wastes versus household waste	[29]
Isolation measures greatly reduced the volume of commercial waste, especially for tourist cities, and part of this waste was transferred to household waste	[30]
With the increase in soap packaging waste production, soap packaging waste management and recycling become essential to reduce environmental impact	[31]
Face masks release microplastics, which are directly inhaled during use or transported through the environment. Additional research is needed on that topic	[38]
The extensive use of personal protective equipment (PPE) driven by the COVID-19 pandemic has become an important contributor to marine plastic pollution	[42]
Lockdowns have led to higher levels of consumption of packaged products, and of take-away food	[43]
The use of disinfectant prior to sorting waste, as well as storing waste for 9 days, may help to inactivate the COVID-19 virus, ensuring an appropriate safety level for MSW management	[48]
The codification of new policies for municipal waste management is necessary	[52]
Waste treatment facilities were overwhelmed, forcing emergency treatment and disposals (e.g., co-disposal in a municipal solid waste incinerator, cement kilns, industrial furnaces, and deep burial) to ramp up processing capacity	[53]
In countries with high recycling rates of MSW, the need to protect MSW employees' health has affected the supply stream of the recycling industry	[54]
The ban on the sale of tobacco products during the lockdown in South Africa did not greatly reduce the number of cigarette butts and associated packaging	[101]

The check of the completeness of Scopus® (or similar databases) is easy to do. Indeed, a recent work accepted in a non-Scopus® international conference [105] reported original data on the fluctuation of quantity and quality of MSW important to understand the effects of the pandemic in regions with a highly efficient source separation. Tables 3 and 4 are reported to see how the first impact of the pandemic acted on MSW in an Italian region where the MSW management is well organized [106], [107]. It can be noticed that:

- lockdown lowered MSW production for contraction of purchases but also for inactivation of eco-centres where to deliver household waste not compatible with curb side collection;
- lockdown lowered selective collection efficiency because of the inactivation of source separation at user level where householders have got COVID-19: there they had to deliver MSW fractions all together until the end of their illness.

These data and comments are not present in Scopus®, neither for similar cases, thus in case of emergency issues it is important to set wider criteria of searching to reach a comprehensive vision of the problem.

Table 3: MSW comparison during lockdown (9th–18th March 2020) and not lockdown (2019) periods in an Italian region [105].

Months	January	February	March	April	May	June
U.M.	(t)	(t)	(t)	(t)	(t)	(t)
2019	13883.58	11397.10	14666.10	14655.94	14866.11	13384.00
2020	14395.44 (+3.69%)	12540.52 (+10.03%)	12278.71 (−16.28%)	11172.56 (−23.77%)	13754.53 (−7.48%)	14510.36 (+8.42%)

Table 4: Percentage of selective collection, relative to the period between January and June of the years 2019 and 2020 for the Trento province [105].

Months	January	February	March	April	May	June
2019	79.82%	79.12%	82.70%	83.22%	83.28%	82.69%
2020	79.99%	80.86%	80.13%	79.93%	83.78%	82.76%

4 CONCLUSIONS

The present paper demonstrated that COVID-19 and MSW is a topic limited to a niche, in spite of the relevance in terms of health management. Moreover, it resulted that the more stimulated scientists belong to the regions with high criticalities during the pandemic. Depending on the short period available to develop deep research, a part of the articles underlines that future research is needed for a complete vision of the problem. International databases resulted important to find information on the topic but they cannot cover all the useful information produced during events as the present pandemic.

REFERENCES

[1] Mahyari, K.F., Sun, Q., Klemes, J.J., Aghbashlo, M., Tabatabaei, M., Khoshnevisan, B. & Birkved M., To what extent do waste management strategies need adaptation to post-COVID-19? *Science of the Total Environment,* **837,** 155829, 2022.

[2] Zajemska, M., Magdziarz, A., Iwaszko, J., Skrzyniarz, M. & Poskart, A., Numerical and experimental analysis of pyrolysis process of RDF containing a high percentage of plastic waste. *Fuel,* **320**, 123981, 2022.

[3] Maiurova, A., Kurniawan, T.A., Kustikova, M., Bykovskaia, E., Othman, M.H.D., Singh, D. & Goh, H.H., Promoting digital transformation in waste collection service and waste recycling in Moscow (Russia): Applying a circular economy paradigm to mitigate climate change impacts on the environment. *Journal of Cleaner Production,* **354**, 131604, 2022.

[4] Mondelli, G., Silva, E.R., Claro, I.C.M., Augusto, M.R., Duran, A.F.A., Cabral, A.D., de Moraes Bomediano Camillo, L., dos Santos Oliveira, L.H. & de Freitas Bueno, R., First case of SARS-CoV-2 RNA detection in municipal solid waste leachate from Brazil. *Science of the Total Environment*, **824**, 15392, 2022.

[5] Kutralam-Muniasamy, G. & Shruti, V.C., The case of "public congregation vs. COVID-19 PPE pollution": Evidence, lessons, and recommendations from the annual pilgrimage to the Catholic Holy Site in Mexico City. Mexico, *Science of the Total Environment*, **821**, 153424, 2022.

[6] Pinto, A.D., Jalloul, H., Nickdoost, N., Sanusi, F., Choi, J., Abichou, T., Challenges and adaptive measures for U.S. municipal solid waste management systems during the COVID-19 pandemic. *Sustainability*, **14**(8), p, 4834, 2022.

[7] Wang, F., Fang, Y., Deng, H. & Wei, F., How community medical facilities can promote resilient community constructions under the background of pandemics. *Indoor and Built Environment*, **31**(4), pp. 1018–1027, 2022.

[8] Shadkam, E., Cuckoo optimization algorithm in reverse logistics: A network design for COVID-19 waste management, *Waste Management and Research,* **40**(4), pp. 458–469, 2022.

[9] Shen, W., Zhu, N., Xi, Y., Huang, J., Li, F., Wu, P. & Dang, Z., Effects of medical waste incineration fly ash on the promotion of heavy metal chlorination volatilization from incineration residues. *Journal of Hazardous Materials*, **425**, 128037, 2022.

[10] Zhao, H., Liu, H., Wei, G., Zhang, N., Qiao, H., Gong, Y., Yu, X., Zhou, J. & Wu, Y., A review on emergency disposal and management of medical waste during the COVID-19 pandemic in China. *Science of the Total Environment*, **810**, 152302, 2022.

[11] Felix, C.B., Ubando, A.T., Chen, W.-H., Goodarzi, V. & Ashokkumar, V., COVID-19 and industrial waste mitigation via thermochemical technologies towards a circular economy: A state-of-the-art review. *Journal of Hazardous Materials,* **423**, 127215, 2022.

[12] Lan, D.-Y., Zhang, H., Wu, T.-W., Lü, F., Shao, L.-M. & He, P.-J., Repercussions of clinical waste co-incineration in municipal solid waste incinerator during COVID-19 pandemic. *Journal of Hazardous Materials,* **423**, 127144, 2022.

[13] Kalina, M., Kwangulero, J., Ali, F. & Tilley, E., "You need to dispose of them somewhere safe": COVID-19, masks, and the pit latrine in Malawi and South Africa. *PLoS ONE,* **17**(2 February), e0262741, 2022.

[14] Dehal, A., Vaidya, A.N. & Kumar, A.R., Biomedical waste generation and management during COVID-19 pandemic in India: Challenges and possible management strategies. *Environmental Science and Pollution Research,* **29**(10), pp. 14830–14845, 2022.

[15] Singh, E., Kumar, A., Mishra, R. & Kumar, S., Solid waste management during COVID-19 pandemic: Recovery techniques and responses. *Chemosphere*, **288**, 132451, 2022.

[16] Ul'yanovskii, N.V., Kosyakov, D.S., Sypalov, S.A., Varsegov, I.S., Shavrina, I.S. & Lebedev, A.T., Antiviral drug Umifenovir (Arbidol) in municipal wastewater during the COVID-19 pandemic: Estimated levels and transformation. *Science of the Total Environment*, **805**, 150380, 2022.

[17] Abonomi, A., De Lacy, T. & Pyke, J., Environmental impact of the Hajj. *International Journal of Religious Tourism and Pilgrimage,* **10**(1), pp. 133–151, 2022.

[18] Ranjbari, M., Shams Esfandabadi, Z., Gautam, S., Ferraris, A. & Scagnelli, S.D., Waste management beyond the COVID-19 pandemic: Bibliometric and text mining analyses. *Gondwana Research*, in press.

[19] [No author name available], International Conference on Advances in Construction Materials and Management, ACMM 2021, *Lecture Notes in Civil Engineering,* **191,** 2022.

[20] Thakur, V., Parida, D.J. & Raj, V., Sustainable municipal solid waste management (MSWM) in the smart cities in Indian context. *International Journal of Productivity and Performance Management*, in press.

[21] Vaidevi, C., Vijayan, D.S., Nivetha, C. & Kalpana, M., COVID-19 future proof infrastructure. *Lecture Notes in Civil Engineering,* **202**, pp. 119–127, 2022.

[22] Ye, Q., Asmi, F., Anwar, M.A., Zhou, R. & Siddiquei, A.N., Health concerns among waste collectors during pandemic crisis. *Environmental Science and Pollution Research,* **29**(5), pp. 6463–6478, 2022.
[23] Anand, U., Li, X., Sunita, K., Lokhandwala, S., Gautam, P., Suresh, S., Sarma, H., Vellingiri, B., Dey, A., Bontempi, E. & Jiang, G., SARS-CoV-2 and other pathogens in municipal wastewater, landfill leachate, and solid waste: A review about virus surveillance, infectivity, and inactivation. *Environmental Research,* **203**, 111839, 2022.
[24] Al-Omran, K., Khan, E., Ali, N. & Bilal, M., Estimation of COVID-19 generated medical waste in the Kingdom of Bahrain. *Science of the Total Environment,* **801**, 149642, 2021.
[25] Sharma, H.B., Vanapalli, K.R., Samal, B., Cheela, V.R.S., Dubey, B.K. & Bhattacharya, J., Circular economy approach in solid waste management system to achieve UN-SDGs: Solutions for post-COVID recovery. *Science of the Total Environment,* **800**, 149605, 2021.
[26] Patrício Silva, A.L., Prata, J.C., Duarte, A.C., Barcelò, D. & Rocha-Santos, T., An urgent call to think globally and act locally on landfill disposable plastics under and after COVID-19 pandemic: Pollution prevention and technological (Bio) remediation solutions. *Chemical Engineering Journal*, **426**, 131201, 2021.
[27] Brohan, M.A., Dom, N.C., Ishak, A.R., Abdullah, S., Salim, H., Ismail, S.N.S. & Precha, N., An analysis on the effect of coronavirus (COVID-19) pandemic movement control order (MCOS) on the solid waste generation in Peninsular Malaysia. *Environmental Science and Pollution Research,* **28**(46), 6pp. 6501–66509, 2021.
[28] Govindan, K., Nasr, A.K., Mostafazadeh, P. & Mina, H., Medical waste management during coronavirus disease 2019 (COVID-19) outbreak: A mathematical programming model. *Computers and Industrial Engineering,* **162**, 107668, 2021.
[29] Cai, M., Guy, C., Héroux, M., Lichtfouse, E. & An, C., The impact of successive COVID-19 lockdowns on people mobility, lockdown efficiency, and municipal solid waste. *Environmental Chemistry Letters,* **19**(6), pp. 3959–3965, 2021.
[30] Liang, Y., Song, Q., Wu, N., Li, J., Zhong, Y. & Zeng, W., Repercussions of COVID-19 pandemic on solid waste generation and management strategies. *Frontiers of Environmental Science and Engineering,* **15**(6), p. 115, 2021.
[31] Chirani, M.R., Kowsari, E., Teymourian, T. & Ramakrishna, S., Environmental impact of increased soap consumption during COVID-19 pandemic: Biodegradable soap production and sustainable packaging. *Science of the Total Environment,* **796**, 149013, 2021.
[32] Mallick, S.K., Pramanik, M., Maity, B., Das, P. & Sahana, M., Plastic waste footprint in the context of COVID-19: Reduction challenges and policy recommendations towards sustainable development goals. *Science of the Total Environment*, **796**, 148951, 2021.
[33] Zhao, H.-L., Wang, L., Liu, F., Liu, H.-Q., Zhang, N. & Zhu, Y.-W., Energy, environment and economy assessment of medical waste disposal technologies in China. *Science of the Total Environment*, **796**, 148964, 2021.
[34] Dutta, S., Environmental treatment technologies for municipal, industrial and medical wastes: Remedial scope and efficacy. *Environmental Treatment Technologies for Municipal, Industrial and Medical Wastes: Remedial Scope and Efficacy*, pp. 1–236, 2021.
[35] Thomine, O., Alizon, S., Boennec, C., Barthelemy, M. & Sofonea, M.T., Emerging dynamics from high-resolution spatial numerical epidemics. *eLife,* **10**, e71417, 2021.

[36] Vu, H.L., Ng, K.T.W., Richter, A., Karimi, N. & Kabir, G., Modeling of municipal waste disposal rates during COVID-19 using separated waste fraction models. *Science of the Total Environment,* **789**, 148024, 2021.

[37] [No author name available], *Proceedings of the 3rd International Conference on Inventive Research in Computing Applications*, ICIRCA, 2021.

[38] De-la-Torre, G.E., Pizarro-Ortega, C.I., Dioses-Salinas, D.C., Ammendolia, J. & Okoffo, E.D., Investigating the current status of COVID-19 related plastics and their potential impact on human health. *Current Opinion in Toxicology*, **27**, pp. 47–53, 2021.

[39] Sebastian, R.M. & Louis, J., Understanding waste management at airports: A study on current practices and challenges based on literature review. *Renewable and Sustainable Energy Reviews,* **147**, 111229, 2021.

[40] Park, C., Choi, H., Andrew Lin, K.-Y., Kwon, E.E. & Lee, J., COVID-19 mask waste to energy via thermochemical pathway: Effect of co-feeding food waste. *Energy,* **230**, 120876, 2021.

[41] Osra, F., Morsy, E.A. & Abd El-Rahim, I.H.A., Guidance plans for solid waste management during COVID-19 in Makkah, Saudi Arabia. *Arabian Journal of Geosciences,* **14**(15), p. 1466, 2021.

[42] Rakib, M.R.J., De-la-Torre, G.E., Pizarro-Ortega, C.I., Dioses-Salinas, D.C. & Al-Nahian, S., Personal protective equipment (PPE) pollution driven by the COVID-19 pandemic in Cox's Bazar, the longest natural beach in the world. *Marine Pollution Bulletin,* **169,** 112497, 2021.

[43] Filho, W.L., Voronova, V., Kloga, M., Paço, A., Minhas, A., Salvia, A.L., Ferreira, C.D. & Sivapalan, S., COVID-19 and waste production in households: A trend analysis. *Science of the Total Environment,* **777**, 145997, 2021.

[44] Yousefi, M., Oskoei, V., Jonidi Jafari, A., Farzadkia, M., Hasham Firooz, M., Abdollahinejad, B. & Torkashvand, J., Municipal solid waste management during COVID-19 pandemic: Effects and repercussions. *Environmental Science and Pollution Research,* **28**(25), pp. 32200–32209, 2021.

[45] Moonsammy, S., Oyedotun, T.D.T., Renn-Moonsammy, D.-M., Oyedotun, T.D., Ally, N., Kasim, O.F. & Famewo, A., COVID-19 effects on municipality waste collection services for households: Statistical modelling of perspectives from Guyana and Nigeria. *Journal of Material Cycles and Waste Management*, **23**(4), pp. 1678–1687, 2021.

[46] Su, G., Ong, H.C., Ibrahim, S., Fattah, I.M.R., Mofijur, M. & Chong, C.T., Valorisation of medical waste through pyrolysis for a cleaner environment: Progress and challenges. *Environmental Pollution,* **279**, 116934, 2021.

[47] Mohamed, K., Amina, M.-S., El Mouaz, M.B., Zihad, B. & Wafa, R., The impact of the coronavirus pandemic on the household waste flow during the containment period. *Environmental Health and Toxicology*, **36**(2), e2021011, 2021.

[48] Das, A.K., Islam, M.N., Billah, M.M. & Sarker, A., COVID-19 and municipal solid waste (MSW) management: A review. *Environmental Science and Pollution Research,* **28**(23), pp. 28993–29008, 2021.

[49] Ganguly, R.K. & Chakraborty, S.K., Integrated approach in municipal solid waste management in COVID-19 pandemic: Perspectives of a developing country like India in a global scenario. *Case Studies in Chemical and Environmental Engineering*, **3**, 100087, 2021.

[50] Pereira de Albuquerque, F., Dhadwal, M., Dastyar, W., Mirsoleimani Azizi, S.M., Karidio, I., Zaman, H. & Dhar, B.R., Fate of disposable face masks in high-solids anaerobic digestion: Experimental observations and review of potential environmental implications. *Case Studies in Chemical and Environmental Engineering*, **3**, 100082, 2021.

[51] Maalouf, A. & Maalouf, H., Impact of COVID-19 pandemic on medical waste management in Lebanon. *Waste Management and Research*, **9**(1), pp. 45–55, 2021.

[52] Torkashvand, J., Jonidi Jafari, A., Godini, K., Kazemi, Z., Kazemi, Z. & Farzadkia, M., Municipal solid waste management during COVID-19 pandemic: A comparison between the current activities and guidelines. *Journal of Environmental Health Science and Engineering*, **19**(1), pp. 173–179, 2021.

[53] Hantoko, D., Li, X., Pariatamby, A., Yoshikawa, K., Horttanainen, M. & Yan, M., Challenges and practices on waste management and disposal during COVID-19 pandemic. *Journal of Environmental Management,* **286**, 112140, 2021.

[54] Vaverková, M.D., Paleologos, E.K., Dominijanni, A., Koda, E., Tang, C.-S., Małgorzata, W., Li, Q., Guarena, N., Mohamed, A.-M.O., Vieira, C.S., Manassero, M., O'Kelly, B.C., Xie, Q., Bo, M.W., Adamcová, D., Podlasek, A., Anand, U.M., Mohammad, A., Goli, V.S.N.S., Kuntikana, G., Palmeira, E.M., Pathak, S. & Singh, D.N., Municipal solid waste management under COVID-19: Challenges and recommendations. *Environmental Geotechnics,* **8**(3), pp. 217–232, 2021.

[55] Zhao, H., Liu, H., Wei, G., Wang, H., Zhu, Y., Zhang, R. & Yang, Y., Comparative life cycle assessment of emergency disposal scenarios for medical waste during the COVID-19 pandemic in China. *Waste Management,* **126**, pp. 388–399, 2021.

[56] Wang, J., Chen, Z., Lang, X., Wang, S., Yang, L., Wu, X., Zhou, X. & Chen, Z., Quantitative evaluation of infectious health care wastes from numbers of confirmed, suspected and out-patients, during COVID-19 pandemic: A case study of Wuhan. *Waste Management,* **126**, pp. 323–330, 2021.

[57] Tsai, W.-T., Analysis of plastic waste reduction and recycling in Taiwan. *Waste Management and Research*, **39**(5), pp. 713–719, 2021.

[58] Karpagam, S. & Dsouza, J., Occupational hazards in healthcare settings a study of invisibilised frontline workers in Bengaluru. *Economic and Political Weekly*, **56**(17), pp. 60–65, 2021.

[59] Roque, A.J., Paleologos, E.K., O'Kelly, B.C., Tang, A.M., Reddy, K.R., Vitone, C., Mohamed, A.-M.O., Koda, E., Goli, V.S.N.S., Vieira, C.S., Fei, X., Sollecito, F., Vaverková, M.D., Plötze, M., Petti, R., Podlasek, A., Puzrin, A.M., Cotecchia, F., Ski, P.O., Mohammad, A., Singh, P., El Gamal, M., Farouk, S., Al Nahyan, M.T., Mickovski, S.B. & Singh, D.N., Sustainable environmental geotechnics practices for a green economy. *Environmental Geotechnics*, **9**(2), pp. 68–84, 2021.

[60] [No author name available], *1st Journal of Environmental Science and Sustainable Development Symposium, IOP Conference Series: Earth and Environmental Science,* **716**(1), 2021.

[61] Rahimi, N.R., Fouladi-Fard, R., Aali, R., Shahryari, A., Rezaali, M., Ghafouri, Y., Ghalhari, M.R., Asadi-Ghalhari, M., Farzinnia, B., Conti Gea, O. & Fiore, M., Bidirectional association between COVID-19 and the environment: A systematic review. *Environmental Research*, **194**, 110692, 2021.

[62] Suthar, S., Das, S., Nagpure, A., Madhurantakam, C., Tiwari, S.B., Gahlot, P. & Tyagi, V.K., Epidemiology and diagnosis, environmental resources quality and socio-economic perspectives for COVID-19 pandemic. *Journal of Environmental Management,* **280**, 111700, 2021.

[63] Fan, Y.V., Jiang, P., Hemzal, M. & Klemeš, J.J., An update of COVID-19 influence on waste management. *Science of the Total Environment*, **754**, 142014, 2021.
[64] Toton, B.N., Eat prosperity campaign: Techno-economic forecasting of a food waste collection scheme for growing economically sustainable algae on landfills, leveraging sustainable infrastructure for resilient communities: Selected papers from the I*nternational Conference on Sustainable Infrastructure 2021*, pp. 1–12, 2021.
[65] Karaeva, A., Conti, F., Torretta, V., Ghiringhelli, G., Magaril, E.R. & Rada, E.C., MSW management in a lacustrine area: Circular economy criteria and effects of COVID-19 emergency. *Proceedings of 2021 10th International Conference on Energy and Environment, CIEM 2021,* Bucharest, Romania, 175003, 2021.
[66] [No author name available], Computing4Human 2021. *Proceedings of the 2nd International Conference on Human-Centered Artificial Intelligence, CEUR Workshop Proceedings,* p. 3026, 2021.
[67] Sapariya, D.D., Patdiwala, U.J., Panchal, H., Ramana, P.V., Makwana, J. & Sadasivuni, K.K., A review on thermochemical biomass gasification techniques for bioenergy production. *Energy Sources, Part A: Recovery, Utilization and Environmental Effects*, in press.
[68] Ventura, K.S., Morais, M.S., Filho, P.V. & Brunetti Junior, A., Analysis of COVID-19 impacts at the collection of municipal household, recyclables and healthcare wastes in the municipality of araraquara (SP), Brazil. *Engenharia Sanitaria e Ambiental*, **26**(4), pp. 775–784, 2021.
[69] Mawkhlieng, U. & Majumdar, A., Waste management of medical personal protective equipment and facemasks: Challenges during and post COVID-19 pandemic. *Environmental Footprints and Eco-Design of Products and Processes*, pp. 37–60, 2021.
[70] Goyal, P. & Pal, A., Solid waste management practices in Indian cities. *Lecture Notes in Mechanical Engineering*, pp. 1143–1155, 2021.
[71] [No author name available], *6th EAI International Conference on Science and Technologies for Smart Cities, SmartCity 2020*, Lecture Notes of the Institute for Computer Sciences, Social-Informatics and Telecommunications Engineering, LNICST, p. 372, 2021.
[72] Pikoń, K., Poranek, N., Czajkowski, A. & Łaźniewska-Piekarczyk, B., Poland's proposal for a safe solution of waste treatment during the COVID-19 pandemic and circular economy connection. *Applied Sciences,* **11**(9), p. 3939, 2021.
[73] Öztürk, I., The COVID-19 pandemic and waste management. *Duzce Medical Journal*, **23**(SI 1), pp. 27–29, 2021.
[74] Suryawan, I.W.K., Rahman, A., Septiariva, I.Y., Suhardono, S. & Wijaya, I.M.W., Life cycle assessment of solid waste generation during and before pandemic of COVID-19 in Bali province. *Journal of Sustainability Science and Management,* **16**(1), pp. 11–21, 2021.
[75] Malkawi, M., Al-Yousfi, B. & Mandil, A., Air quality and health impacts in the eastern Mediterranean region: An eye on COVID-19. *Eastern Mediterranean Health Journal,* **27**(1), pp. 3–4, 2021.
[76] Zand, A.D. & Heir, A.V., Emanating challenges in urban and healthcare waste management in Isfahan, Iran after the outbreak of COVID-19. *Environmental Technology,* **42**(2), pp. 329–336, 2021.

[77] Karunanidhi, D., Aravinthasamy, P., Deepali, M., Subramani, T. & Shankar, K., Groundwater pollution and human health risks in an industrialized region of southern India: Impacts of the COVID-19 lockdown and the monsoon seasonal cycles. *Archives of Environmental Contamination and Toxicology*, **80**(1), pp. 259–276, 2021.

[78] Mostafa, M.K., Gamal, G. & Wafiq, A., The impact of COVID 19 on air pollution levels and other environmental indicators: A case study of Egypt. *Journal of Environmental Management,* **277**, 111496, 2021.

[79] Zand, A.D. & Heir, A.V., Environmental impacts of new Coronavirus outbreak in Iran with an emphasis on waste management sector. *Journal of Material Cycles and Waste Management,* **23**(1), pp. 240–247, 2021.

[80] Mohammad, A., Goli, V.S.N.S. & Singh, D.N., Discussion on "Challenges, opportunities, and innovations for effective solid waste management during and post COVID-19 pandemic, by Sharma et al. (2020)". *Resources, Conservation and Recycling*, **164**, 105175, 2021.

[81] Penteado, C.S.G. & Castro, M.A.S.D., COVID-19 effects on municipal solid waste management: What can effectively be done in the Brazilian scenario? *Resources, Conservation and Recycling*, **164**, 105152, 2021.

[82] Yang, L., Yu, X., Wu, X., Wang, J., Yan, X., Jiang, S. & Chen, Z., Emergency response to the explosive growth of health care wastes during COVID-19 pandemic in Wuhan, China. *Resources, Conservation and Recycling,* **164**, 105074, 2021.

[83] Ragazzi, M., Rada, E.C. & Schiavon, M., Municipal solid waste management during the SARS-COV-2 outbreak and lockdown ease: Lessons from Italy. *Science of the Total Environment,* **745**, 141159, 2020.

[84] Kulkarni, B.N. & Anantharama, V., Repercussions of COVID-19 pandemic on municipal solid waste management: Challenges and opportunities. *Science of the Total Environment*, **743**, 140693, 2020.

[85] Naughton, C.C., Will the COVID-19 pandemic change waste generation and composition? The need for more real-time waste management data and systems thinking. *Resources, Conservation and Recycling,* **162**, 105050, 2020.

[86] Zand, A.D. & Heir, A.V., Emerging challenges in urban waste management in Tehran, Iran during the COVID-19 pandemic. *Resources, Conservation and Recycling,* **162**, 105051, 2020.

[87] Zaleski, P. & Chawla, Y., Circular economy in Poland: Profitability analysis for two methods of waste processing in small municipalities. *Energies*, **13**(19), p. 5166, 2020.

[88] Lakhouit, A., Practical recommendations for temporary storage of medical wastes during the COVID-19 pandemic. *Indoor and Built Environment*, **29**(8), pp. 1186–1188, 2020.

[89] Mihai, F.-C., Assessment of COVID-19 waste flows during the emergency state in Romania and related public health and environmental concerns. *International Journal of Environmental Research and Public Health,* **17**(15), pp. 1–18, 2020.

[90] Ma, Y., Lin, X., Wu, A., Huang, Q., Li, X. & Yan, J., Suggested guidelines for emergency treatment of medical waste during COVID-19: Chinese experience. *Waste Disposal and Sustainable Energy*, **2**(2), pp. 81–84, 2020.

[91] Rada, E.C., Magaril, E.R., Schiavon, M., Karaeva, A., Chashchin, M. & Torretta, V., MSW management in universities: Sharing best practices. *Sustainability,* **12**(12), p. 5084, 2020.

[92] Oluwalana, A., Etu, E.-E. & Tenebe, T., A concise review on municipal solid waste management in a pandemic era: Knowledge gaps identified for developed and developing countries. *Proceedings of the International Conference on Industrial Engineering and Operations Management*, 2020.

[93] Septiariva, I.Y., Suryawan, I.W.K., Sarwono, A. & Ramadan, B.S., Municipal infectious waste during COVID-19 pandemic: Trends, impacts, and management. *International Journal of Public Health Science,* **11**(2), pp. 552–557, 2022.

[94] Santos, A.A., da Silva, A.F., Gouveia, A., Caetano, N. & Felgueiras, C., Recyclable waste collection: Increasing ecopoint filling capacity to reduce energy for transportation. *Energy Reports,* **8**, pp. 430–436, 2022.

[95] Valizadeh, J. & Mozafari, P., A novel cooperative model in the collection of infectious waste in COVID-19 pandemic. *Journal of Modelling in Management*, **17**(1), pp. 363–401, 2022.

[96] Valizadeh, J., Aghdamigargari, M., Jamali, A., Aickelin, U., Mohammadi, S., Khorshidi, H.A. & Hafezalkotob, A., A hybrid mathematical modelling approach for energy generation from hazardous waste during the COVID-19 pandemic. *Journal of Cleaner Production*, **315**, 128157, 2021.

[97] Prats, R.M., van Drooge, B.L., Fernández, P., Marco, E. & Grimalt, J.O., Changes in urban gas-phase persistent organic pollutants during the COVID-19 lockdown in Barcelona. *Frontiers in Environmental Science*, **9**, 650539, 2021.

[98] Zhang, Y., Liu, X., Fang, Y., Liu, D., Tang, A. & Collett, J.L., Atmospheric ammonia in Beijing during the COVID-19 outbreak: Concentrations, sources, and implications. *Environmental Science and Technology Letters*, **8**(1), pp. 32–38, 2021.

[99] Zand, A.D. & Heir, A.V., Emanating challenges in urban and healthcare waste management in Isfahan, Iran after the outbreak of COVID-19. *Environmental Technology*, **42**(2), pp. 329–336, 2021.

[100] Zand, A.D. & Heir, A.V., Environmental impacts of new Coronavirus outbreak in Iran with an emphasis on waste management sector. *Journal of Material Cycles and Waste Management*, **23**(1), pp. 240–247, 2021.

[101] Ryan, P.G., Maclean, K. & Weideman, E.A., The impact of the COVID-19 lockdown on urban street litter in South Africa. *Environmental Processes*, 7(4), pp. 1303–1312, 2020.

[102] Zand, A.D. & Heir, A.V., Emerging challenges in urban waste management in Tehran, Iran during the COVID-19 pandemic. *Resources, Conservation and Recycling,* **162**, 105051, 2020.

[103] Corrêa, H.L. & Corrêa, D.G., Polymer applications for medical care in the COVID-19 pandemic crisis: Will we still speak ill of these materials? *Frontiers in Materials*, **7**, p. 283, 2020.

[104] European Commission, Waste management in the context of the coronavirus crisis, 2020. https://ec.europa.eu/info/sites/default/files/waste_management_guidance_dg-env.pdf. Accessed on: 7 May 2022.

[105] Malesardi, N., Ragazzi, M. & Rada, E.C., Municipal solid waste management during the pandemic in EU contest: First COVID-19 wave. *International Conference SUM 2022*, Capri, Italy, 2022.

[106] Rada, E.C., Zatelli, C., Cioca, L.I. & Torretta, V., Selective collection quality index for municipal solid waste management. *Sustainability*, **10**(1), p. 257, 2018.

[107] Rada, E.C., Zatelli, C. & Mattolin, P., Municipal solid waste selective collection and tourism. *WIT Transactions on Ecology and the Environment*, vol. 180, WIT Press: Southampton and Boston, pp. 187–197, 2014.

BIOMEDICAL SOLID WASTE MANAGEMENT IN LOW-INCOME CONTEXTS: THE CASE OF BEIRA, MOZAMBIQUE

ADA CASTELLUCCI[1,2], MARCO RAGAZZI[1], FEDERICO BERGHI[3], DARIO GUIRRERI[3] & ISACCO RAMA[2]
[1]Environmental and Mechanical Engineering, University of Trento, Italy
[2]Consortium of Associations with Mozambique Onlus (CAM), Italy
[3]Progettomondo NGO, Italy

ABSTRACT
The lack of adequate waste management practices is one of the most critical and common challenges in a low-income urban context. In this sense, the environmental release of biomedical solid waste without proper treatment is a serious risk for the city of Beira in Mozambique. The present work is developed in the frame of the LimpaMOS MOÇambique project (co-financed mainly by the Italian Agency for Development Cooperation) and SIRSU project (Italian Agency for Development Cooperation). These projects involve Italian and Mozambican actors from the public, private, and non-profit sectors. One of the aims of the projects is to strengthen the proper management, separation, and disposal of biomedical solid waste. The present study investigates two main aspects: (a) the separation and collection of hospital waste in the health units distributed throughout the city; and (b) the centralised treatment in the Beira Central Hospital carried out by an incinerator. Concerning the first point, the case study of the Ponta-Gêa Health Centre shows how it is possible to improve waste sorting, which is currently problematic mainly due to the lack of plastic bags, central for a safe collection and transport but not economically nor environmentally sustainable. Concerning the second point, the Beira Central Hospital incinerator presents significant problems in terms of environmental impact, health, and safety. Using weighing, estimates, and official documentation from the Municipality of Beira, the total production of biomedical solid waste in the city was assessed. The short-term aim of this study is an efficient distribution of the biomedical solid waste between the incinerator and a new electric sterilizer. Finally, the possibility of opening new and more sustainable plants is being evaluated, with the long-term aim of closing the incinerator.
Keywords: biomedical solid waste, hospital waste, low-income countries, Mozambique, management, segregation, incinerator, sterilizer .

1 INTRODUCTION

Of the total amount of waste generated by healthcare activities, about 85% is general, non-hazardous waste. The remaining 15% is the so-called biomedical solid waste (BSW) and it is considered hazardous material that may be infectious, toxic, or radioactive [1]. Low-income countries generate on average 0.2 kg of BSW per hospital bed per day, 0.3 kg less than high-income countries [1]. However, without good practices for healthcare waste separation and management, the effective amount of BSW mixed with all kinds of hospital solid waste (HSW) becomes much higher, thus increasing the possibility of spreading infections [2]–[4].

Mozambique ranks low on the human development index (HDI) at 0.456, or 181 out of 189 countries [5]. Since its recent independence from Portugal in 1975, Mozambique has been affected by numerous natural disasters. Violent tropical cyclones like Idai (Beira, March 2019) or Eloise (Beira, January 2021) strike a vulnerable population, creating circumstances that exacerbated poverty levels, food insecurity, and spread of diseases among the affected people [6]. In this complex and fragile climatic, historical, social, and economic context, a deep crisis is affecting the entire public health sector, where the already precarious economic

www.witpress.com, ISSN 1743-3541 (on-line)
doi:10.2495/WMEI220091

conditions are further aggravated by onerous interventions following the various natural disasters and the COVID-19 emergency [7]–[9]. Correct and safe collection of biomedical solid waste is a frequently underestimated aspect of hospital management, aggravated by weaknesses in collection, transport, and final disposal.

Beira is the capital and largest city of Sofala Province in central Mozambique and has a population of 673,685 [10]. In an area of 633 km^2 there are 16 health units of the National Health Service, including a central hospital, a provincial hospital, 12 health centres and two health posts, plus several private clinics. 1,779 people are employed as health personnel, of which 745 are in the Central Hospital, which is the largest in the city [10].

This study aims to lay the foundations for an improvement in the management of HSW in Beira. To do it, the starting point was to analyse the current municipal solid waste (MSW) and biomedical solid waste management system and also the differences between the various health units in the city. MSW management is at the beginning stage and most of this waste is disposed in open dumps [11]–[13]. This is a normal practice in low-income and developing countries [13]–[16].

The health centres are halfway between small hospitals and doctors' surgeries. They do not have the capacity to admit patients, except for maternity wards in only a few of the larger health centres, which are able to admit women for up to 72 hours. Only a few departments, such as the maternity ward and the emergency room, operate on a 24-hour basis, but they are not organised to deal with complications or surgeries, which can only be managed at Beira Central Hospital (the biggest one in the city). Consequently, even the management of the HSW differs according to the type and the size of the health unit. Some health centres are included in waste collection and transport routes, which are followed by centralised treatment (an incinerator) at the Central Hospital, others treat it locally, others do not treat it at all. This study focuses on Ponta-Gêa Health Centre (one of the biggest health centre in the city) and the Central Hospital. Proper waste sorting in health centres is essential for its subsequent correct treatment. Incineration, sterilisation, and dumping will be the three possible BSW management solutions in Beira in the coming years, and each cannot be disconnected from the initial waste sorting.

2 MATERIALS AND METHODS

The Mozambique Environment Programme started in 2018 thanks to the two Italian NGOs "CAM Consorzio Associazioni con il Mozambico" and "ProgettoMondo MLAL" and the municipalities of Beira and Nampula. This 5-year programme operates in a rapidly evolving context of municipal solid waste management, whose problems are many and varied, due to the quantity, variety and distribution of waste produced. Funded mainly by the Autonomous Province of Trento, the European Union Delegation in Maputo and the Italian Agency for Development Cooperation, this programme includes the LimpaMOS MOÇambique and SIRSU projects. Both projects are still ongoing.

One of the many objectives of the LimpaMOS MOÇambique project is to strengthen the proper management, segregation, and disposal of the BWS in at least two urban health units. The LimpaMOS MOÇambique team produced important research on HSW management in the Ponta-Gêa Health Centre. In 2020, they analysed the HSW produced and the results of the 9-day waste weighing indicate an average daily waste production of 55.7 kg. Of this, 32.6 kg is BSW of which 83% is infected waste, 8% is sharps waste, and 9% is anatomical waste. After discovering the daily waste streams and the internal management of HSW, one of the actions of the project involves the construction of a temporary storage facility for HSW, separate from the other buildings of the centre and protected from atmospheric events and unwanted access.

The SIRSU project (Local Enterprise Development for Solid Waste Disposal) involves the creation of a service company specialising in the collection of BSW, which not only aims to protect citizens of Beira and the environment by properly treating hazardous waste, but also to empower local people by creating specialised staff. By donating an NW15 electric sterilizer from the Italian company Newster Group, the project will make an active contribution to the disposal of BSW from the Central Hospital, the public health centres, and at least from five private clinics, and will reduce the total amount of waste sent to the incinerator installed in the Central Hospital.

Working with Beira Municipal Council, CAM developed and improved the new Hospital Solid Waste Collection Plan for the collection and transport of HSW from 15 urban health centres of the National Public Service and four private clinics. This document also contains estimated date of HSW streams produced by health centres. Through the analysis of the collection routes and field visits, it was verified that the entire HSW management chain in the city of Beira is critical, starting from the first sorting in hospital rooms to the final disposal.

Unlike what is stated in the Hospital Solid Waste Collection Plan, Beira Municipal Council oversees HSW collection and transport only in nine health centres and in five private clinics. all the health centres should be required to implement their waste sorting, but this is not happening both because of a deep crisis in the health sector and because of a deficiency in risk perception. HSW is not always correctly sorted into the categories of infectious, anatomical, sharps, pharmaceutical (the four categories of BSW) and common. Collection and transport carried out by the municipal operators are very inadequate: the vehicle is a tractor with a flatbed trailer, loaded manually, and the waste collectors are not equipped with all the personal protection equipment (PPE). When waste is unpacked and mixed in the temporary storage containers of a health centre, these are emptied directly into the tractor trailer. Even if waste is correctly sorted and packaged (plastic bags containing infective, anatomical, or common waste and carton boxes containing sharps waste) it is not handled safely, and several operators have already been involved in accidents involving needles. In general, the entire transport system does not always guarantee either the separation of the waste nor the safety of the operators and of the community.

The final destinations for HSW are the Central Hospital incinerator for BSW, and the dumpsite for common waste. The incinerator is currently (February 2022) the only BSW treatment site for the city of Beira. Designed to incinerate only the BSW from the Central Hospital by working 3 days a week, the incinerator is used every day except Sunday to burn waste from the majority of the city's health centres, both public and private. Such frequent operation is necessary due to the large quantities of waste and because the cold storage room has not been in operation for about 2 years. The plant uses obsolete technology, resulting in a very high environmental impact. The waste is manually loaded by the operator into a combustion chamber without the closing door. There are two engines for powering the incinerator. The first and only one currently in operation drives a fan that sprays fuel into the combustion chamber, contributing to the increase in temperature for load cycles with high humidity (mainly due to the presence of anatomical waste). The second engine, which has been out of use for several years, is the one used to operate the filter for the smoke generated by combustion. As there are no filters, the only environmental measure should be the chimney, which is too short to provide any real benefits (less than 10 m). Residents living close to the hospital often complain about the smoke from the chimney and the bad smell and are consequently unable to open their windows all morning. The incinerator has other technical problems. The room where it is located has a sheet metal roof, which is often removed by wind during the rainy season. Water falls on the plant and, over time, the outer metal layer has corroded and even has holes in the top. The protective layer inside the combustion chamber is also deteriorating badly, deforming the walls themselves.

Once the main criticalities of the HSW management chain were identified, a second phase of field investigation and subsequent analyses was carried out, due to the absence of data or information. First, it was not clear why the waste on the tractor trailer of the municipal collection and transport service arrived mixed and unpacked at the incinerator, thus denying the operators the possibility to unload and incinerate it. This waste therefore follows the final deposition of the common fraction of HSW and is transported to the dump, despite its hazardous nature. Second, it was not clear how much waste the health units and the Central Hospital produce on average each day.

3 RESULTS

Although a more technologically advanced treatment plant for BSW is indispensable in Beira, first a good waste sorting at the base of the chain of HSW management is needed. This makes it possible to load the two plants correctly, avoiding energy and economic waste, with the aim of reducing the current high environmental impact and the risk to the workers and to the entire community.

The visits and discussions with waste management staff revealed the main criticalities of the system of collection, storage, and transport of HSW in Ponta-Gêa Health Centre, and are listed below:

- Many departments of the hospital complain of a lack of bins, which are therefore insufficient to cover all classes of waste to be sorted;
- The health centre often suffers from a severe lack of disposable plastic bags to collect and separate common, anatomical, and infectious waste. On such occasions, the few available are used primarily for anatomical waste, and only occasionally for infectious waste;
- The wards are not efficiently organised: some rooms have not enough bins, others have many but not clearly classified. This makes collection by waste management staff difficult;
- Many doctors and nurses are not trained in how to carry out efficient waste sorting, which has a very negative effect on the quality of separation;
- Common waste should be taken to the appropriate container outside the health centre and be collected by municipal solid waste service, but the container isn't always available;
- The location for the temporary storage of the BSW was not appropriate before the construction of the new building by LimpaMOS project. It was an open space with a few 240 litres containers, some with lids and wheels and some without. Waste was therefore exposed to weather phenomena and unauthorised access (vandalism, animal attraction, children attraction);
- The municipal operators who empty the containers into the tractor trailer to transport them to the incinerator have little PPE and in very bad conditions;
- The tractor is not a good vehicle for transporting hazardous waste;
- In general, waste sorting is not efficiently carried out. Very often hazardous waste is mixed with common waste and ends up in dumps without any treatment.

To improve waste sorting three potential scenarios were discussed with the chief of waste management staff. The common waste produced represents almost 42% of the total and it is essential that in all scenarios it remains separated from the BSW.

1. **Adequate supply of plastic bags.** The Service for Health, Woman and Social Action of Beira District (SDSMAS) is the institution to which the health centres refer for the supply of new plastic bags. Often the delivery of these materials, as well as of PPE, does not take place or only in very small quantities. For its part, the health centre could start recording the average consumption of bags over a given period and they must ensure that plastic bags are used correctly, without theft or misuse. Due to the crisis in the health sector, there is a tendency not to give proper economic and organisational importance to waste management, but with better budgeting the District Service could improve the supply of plastic bags.
2. **"Plastic-free" management.** The tractor picks up the 240 litre containers from the health centre's temporary storage facility filled with pre-sorted waste. Anatomical waste is the only waste that still needs a plastic bag. By transporting the containers, the waste is not loose and does not risk mixing between different classes. The tractor, loaded with the containers, arrives at the Central Hospital. There, once the new sterilizer is in operation, the containers of infectious waste are emptied directly into it.
3. **Intermediate solution.** An intermediate solution between the previous two is to place 240 litre containers in each department, one for common waste and one for infectious waste. Only the container for infectious waste has a large plastic bag. Each room or examination room has its own 10 or 20 litre bins (three or two depending on the production of anatomical waste) which can be left without a bag (except for the anatomical waste bins) and washed when emptied. They are emptied as needed or at least once a day by the collection staff into the 240 litre containers located at a strategic point on each floor or department. Only at the end of the day the plastic bag is closed and taken to the temporary storage.

To have a better knowledge about waste streams, every day for two weeks the quantities of anatomical, sharps, infectious and pharmaceutical waste were measured, divided by origin: Beira Central Hospital, health units subject to Municipal Council collection, private clinics transporting their own BSW. The daily fuel consumption of the incinerator was also tracked. Table 1 shows the average daily quantities of BSW, while Table 2 shows the incinerator's consumption. In Fig. 1 more data about the average daily amounts and percentages of BSW treated by the incinerator are reported.

The weighing activity showed a significant deficit of HSW compared to previous reports (HSW Collection Plan, an internal CAM report on the initial information gathered at the Central Hospital). The reasons may be either a reduction of patients due to the COVID-19 pandemic or the inability to weigh and treat part of the BSW as it is not packed. For these

Table 1: Average daily amounts of BSW treated by the incinerator sorted by type and place of production: Beira Central Hospital (BCH), health centres (HC), private clinics (PC).

Type of waste	Place of production			
	BCH	HC	PC	TOTAL
Anatomical	20.08 kg	28.75 kg	1 kg	49.83 kg
Sharps	20.42 kg	8.67 kg	0.66 kg	29.75 kg
Infectious	223 kg	7 kg	23.67 kg	253.67 kg
Pharmaceutical	26.08 kg	0 kg	0 kg	26.08 kg
TOTAL	289.58 kg	44.42 kg	25.33 kg	359.33 kg

Table 2: Current and average incinerator consumption.

Type	Quantity
Total amount of biomedical solid waste weighed	4,312 kg
Total amount of fuel used	225 l
Number of days for weighing	12 days
Average diary amount of fuel used	18.75 l
Average monthly amount of fuel used	487.5 l
Average amount of fuel used to burn 1 kg of waste	0.052 l

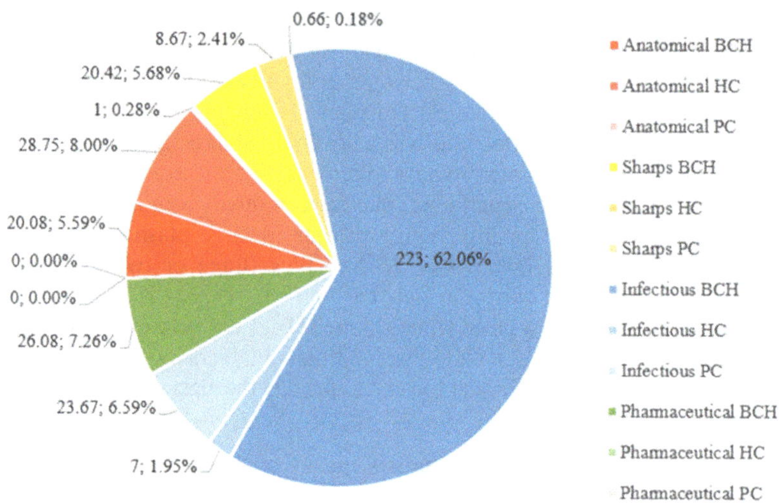

Figure 1: Average daily amounts and percentages of BSW treated by the incinerator expressed in kg, sorted by type and place of production (Beira Central Hospital (BCH), health centres (HC), private clinics(PC)).

reasons, and because of the inaccuracies that could arise from a short weighing period (only two weeks), it would be ill-advised to use the data in Table 1 to assess the load of the two plants. This consideration also stems from an optimistic scenario of an improved collection and transport system. Currently, most of the waste that remains in the tractor trailer as unpackaged (and therefore delivered to the dump) is infectious waste, estimated at 40 kg per day. The average daily amount of BSW produced at the three different sites is reported to be around 550 kg, so the final quantities by category are estimated as in Table 3.

Table 3: Estimate of the average daily production of BSW in the city of Beira, sorted by type.

Type of waste	Weighed and estimated	Percentage	Average daily production
Anatomical	50 kg	12.5%	69 kg
Sharps	30 kg	7.5%	41 kg
Infectious	294 kg	73.5%	404 kg
Pharmaceutical	26 kg	6.5%	36 kg
TOTAL	400 kg	100%	550 kg

NW15 sterilizer can process a maximum of 30 kg of BSW per cycle, which lasts between 30 and 50 minutes. Considering an optimistic scenario in which the machine operators work two seven-hour shifts each, the number of daily cycles can be estimated at 14 and the quantities treated daily at about 420 kg, less than the total 550 kg. Furthermore, to allow optimal treatment and delay the deterioration time of the sterilizer, it is not allowed to be loaded with all BSW fractions, and many types of waste have maximum weight percentages for each load. First of all, the inclusion of even small amounts of common waste (especially organic waste) must be avoided. Ordinary waste should not arrive at the two plants, but if it does because of inaccuracies in separate collection, it is essential to direct it to the incinerator. Thus, it would be unwise to think of treating all BSW exclusively with the sterilizer and not also turning on the incinerator. The main considerations on flow diversion with regard to different types of waste are given below.

- **Anatomical waste.** The large amount of moisture contained in the anatomical waste is a major problem for both types of plant. In the incinerator, the anatomical waste requires large additions of fuel to be incinerated; in the sterilizer this amount must be below 25%–30% of the total load. The amount of anatomical waste is 69 kg, which represents about 16.4% of the total 420 kg that can be treated daily. This is a significant achievement, as it would considerably reduce the use of fuel for burning in the incinerator;
- **Sharps waste.** The sterilizer can process spent syringes that the incinerator cannot destroy. The only caution is how to load the carton boxes containing the syringes: it is advisable to place them at the bottom of the sterilisation chamber, near the blades;
- **Pharmaceutical waste.** Similar to common waste as it is mainly composed of paper boxes, blisters, or in general paper and/or plastic material, it is not recommended to be treated in the steriliser. As it is a material that burns well it can continue to be incinerated;
- **Glass medicine bottles.** This pharmaceutical waste is currently problematic, as incinerator operators have to shred them manually before placing them in the incinerator. Smaller quantities of glass can instead be processed in the sterilizer, thus avoiding further risks to operators.

4 DISCUSSION AND CONCLUSION

The previous paragraph shows that the use of the sterilizer will significantly reduce the load on the incinerator. However, this is a good objective in the short to medium term. The primary long-term objective is necessarily to close the incinerator, a worn-out and problematic plant, but the start-up of a single NW15 steriliser does not allow this objective to be achieved. Considering the weighing carried out and a daily production of about 360–400 kg of BSW, hypothetically the NW15 would be able to treat all the city waste every day. But because of the pandemic period, which discourages patients from going to hospitals, the continuous increase in population and the possibility of breakdown and/or maintenance of the steriliser, it is not prudent to consider that all BSW could be treated with the steriliser in the short future. In this sense, its first year of use will be crucial to confirm or disprove this assumption.

The possibility of closing the incinerator in Beira and some solutions are reported below. However, each proposal will require feasibility studies, interviews, and in-depth research. Studying the city and its current waste management, the following options were considered:

1. **Increasing the operating shifts of NW15 from two (14 hours) to three (21 hours).** The change to three shifts, each of 7 hours, would ensure the daily processing of approximately 630 kg of BSW, thus covering the planned total of 550 kg. This solution would involve a night shift and the employment of at least one other person but would have the great advantage of not having to purchase any additional plant. The main

problem is the centralisation of processing, which would continue to be carried out in one place. This would violate one of the key principles of waste disposal, which is to treat waste as close as possible to its place of production.

2. **Starting up a NW5 steriliser.** The Newster5 sterilizer is an equivalent machine to the NW15, with the difference that its load does not exceed 15 kg per cycle. This means that the sterilizer is capable of processing 210 kg of BSW (over two shifts, totalling 14 hours), which, when added to the 420 kg of the NW15, would cover the total planned quantity. The advantage of this proposal is that it would be possible to decentralise the BSW treatment site, by choosing to locate NW5 in one of the city's major health centres, so that it would treat the waste produced there every day, as well as waste from some neighbouring clinics. This would involve the creation of new waste collection routes, one for the Central Hospital and one for the new health centre where the plant would be located.
3. **Starting up an autoclave.** A similar argument can be made for an autoclave to the NW5 sterilizer. This machine can treat about 50–70 kg in a 35-minute cycle [17]. Considering loading, unloading, and cleaning operations, the indicative treatment of 50 kg per hour can be assumed, ensuring total coverage of the daily BSW quantities. However, the effectiveness of autoclaving as a method of treating hospital waste has not been fully demonstrated, and its use is preferred for the sterilisation of medical instruments [18].
4. **Starting up small incinerators.** Not an environmentally sustainable solution, the installation of small furnaces for incinerating BSW are very common in developing countries. Compared to the current incinerator, small furnaces would have a much smaller load, leading to lower and decentralised emissions. If well designed, with a good filter and sufficient chimney height, small incinerators can be an acceptable solution, considering the context, given the lower costs compared to a sterilisation plant. Of course, an adequate management of residues must be guaranteed.

For both the Ponta-Gêa Health Centre and the evaluation of new waste treatment plants, it is difficult to define an optimal solution. The choices are complicated by a wide fragmentation of management responsibilities and a severe economic crisis in the health sector. However, health centres in Beira already prioritise the use of plastic bags for waste packaging in certain departments or for certain fractions, and the intermediate solution of reducing their number in everyday use could be a valid effort to implement waste collection without a high economic impact. Assessing which new plant could permanently replace the incinerator is inseparable from a review of the entire Hospital Solid Waste Collection Plan.

The increase in NW15 shifts, although the best solution in terms of emissions and costs, has the double disadvantage of the absence of plants running in parallel (essential in case of breakdowns) and the maintenance of very long transport routes for the BSW collected. Of the decentralisation proposals, the inclusion of an NW5 could be more effective in terms of bacterial removal than an autoclave, but not necessarily less expensive.

The construction of small incineration furnaces is the option that should be discouraged the most, but at the same time it may prove to be the one with the fewest economic and acceptance obstacles. Further analysis, availability of technical expertise and economic availability will be the key discriminators in understanding which project proposal is the most suitable for the Beira context.

The safe and environmentally friendly management of the BSW is an issue that could become more relevant after the consolidation of the use of the sterilizer, when the benefits of reducing the use of the incinerator will be more evident to both hospital technicians and administrators.

ACKNOWLEDGEMENTS

This paper was written within the project "Limpamos Moçambique" (AID code 011891), implemented with the contribution of the Italian Agency for Development Cooperation (AICS). The responsibility for the content of this publication lies with the authors and the project partners and does not necessarily represent the views of the AICS funding body.

REFERENCES

[1] World Health Organization, Health-care waste. www.who.int/news-room/fact-sheets/detail/health-care-waste. Accessed on: 14 Dec. 2021.

[2] Ogunsola, F.T. & Mehtar, S., Challenges regarding the control of environmental sources of contamination in healthcare settings in low-and middle-income countries: A narrative review. *Antimicrobial Resistance and Infection Control*, **9**(1), p. 81, 2021.

[3] Kyomba, G.K., Konde, J.N.N., Saila-Ngita, D., Solo, T.K. & Kiyombo, G.M., Assessing the management of healthcare waste for disease prevention and environment protection at selected hospitals in Kinshasa, Democratic Republic of Congo. *Waste Management and Research*, **39**(10), pp. 1237–1244, 2021.

[4] Johnson, K.M., González, M.L., Dueñas, L., Gamero, M., Relyea, G., Luque, L.E. & Caniza, M.A., Improving waste segregation while reducing costs in a tertiary-care hospital in a lower-middle-income country in Central America. *Waste Management and Research*, **31**(7), pp. 733–738, 2013.

[5] United Nations Development Programme, Human Development Index (HDI). hdr.undp.org/en/indicators/137506. Accessed on: 14 Dec. 2021.

[6] Inter-Agency Humanitarian Evaluation Steering Group, Inter-agency humanitarian evaluation of the response to Cyclone Idai in Mozambique. www.wfp.org/publications/inter-agency-humanitarian-evaluation-response-cyclone-idai-mozambique. Accessed on: 14 Dec. 2021.

[7] Langa, E.S., Massuanganhe, J.A. & Nhanala, G.A., The impact of the coronavirus (COVID-19) and climate change on the exchange rate: A multivariate approach for Mozambique. *Prometeica*, **24**, pp. 210–226, 2022.

[8] Salvucci, V. & Tarp, F., Poverty and vulnerability in Mozambique: An analysis of dynamics and correlates in light of the COVID-19 crisis using synthetic panels. *Review of Development Economics*, **25**(4), pp. 1895–1918, 2021.

[9] Mugabe, V.A., Gudo, E.S., Inlamea, O.F., Kitron, U. & Ribeiro, G.S., Natural disasters, population displacement and health emergencies: Multiple public health threats in Mozambique. *BMJ Global Health*, **6**(9), e006778, 2021.

[10] Instituto Nacional de Estatística, Delegação Provincial de Sofala, Anuário Estatístico 2020 Província de Sofala. www.ine.gov.mz/estatisticas/publicacoes/anuario/anuario-provincia-de-sofala/ae-sofala-2020.pdf/view. Accessed on: 15 Dec. 2021.

[11] Tas, A. & Belon, A., *A Comprehensive Review of the Municipal Solid Waste Sector in Mozambique*, Carbon Africa Limited: Nairobi, pp. 1–80, 2014.

[12] Sallwey, J., Hettiarachchi, H. & Hülsmann, S., Challenges and opportunities in municipal solid waste management in Mozambique: A review in the light of nexus thinking. *AIMS Environmental Science*, **4**(5), pp. 621–639, 2017.

[13] Guirreri, D., Silvestri, S., Cristoforetti, A., Rama, I., Giurea, R., Magaril, E. & Rada, E.C., Composting strategy for developing cities: A case study of Beira, Mozambique. *WIT Transactions on Ecology and the Environment*, vol. 247, WIT Press: Southampton and Boston, pp. 69–79, 2020.

[14] Mmereki, D., Current status of waste management in Botswana: A mini-review. *Waste Management and Research*, **36**(7), pp. 555–576, 2018.

[15] Ferronato, N., Torretta, V., Ragazzi, M. & Rada, E.C., Waste mismanagement in developing countries: A case study of environmental contamination. *UPB Scientific Bulletin, Series D: Mechanical Engineering*, **79**(3), pp. 185–196, 2017.
[16] Henry, R.K., Yongsheng, Z. & Jun, D., Municipal solid waste management challenges in developing countries: Kenyan case study. *Waste Management*, **26**(1), pp. 92–100, 2006.
[17] Medical Expo, Gient Medical waste autoclave MWI-500. www.medicalexpo.it/prod/gient/product-95343-822317.html.
[18] Hossain, S., Balakrishnan, V., Ab Rahman, N.N.N., Sarker, Z.I. & Ab Kadir, M.O., Treatment of clinical solid waste using a steam autoclave as a possible alternative technology to incineration. *International Journal of Environmental Research and Public Health*, **9**(3), pp. 855–867, 2012.

IN-DEPTH ANALYSIS ON ODOUR DISPERSION MODELLING AND ITS APPLICATIONS TO WASTE MANAGEMENT OPERATIONS

LUCA ADAMI, MARCO SCHIAVON & MARCO TUBINO
Department of Civil, Environmental and Mechanical Engineering, University of Trento, Italy

ABSTRACT

The management of waste entails the emission of a large variety of compounds into the atmosphere. Waste management processes (e.g. collection and transportation of waste, discharging of waste in dedicated facilities, mechanical–biological treatments and landfill disposal) are known for causing the problem of odour nuisance in the vicinity of waste treatment plants. Substances like volatile organic compounds, sulphides and nitrogen-based compounds are usually associated with waste management processes and, in general, are characterised by low odour threshold values, i.e. they are detected by the human sense of smell even at relatively low concentrations in ambient air. Dispersion modelling represents a fundamental step for the estimation of the odour impact near odour emission sources. However, the results of odour dispersion simulations are strongly affected by the initial hypotheses on the emission sources considered, by specific modelling parameters and by the quality of meteorological and morphological input data. In addition, the variability in the human perception of odour may not allow making universal conclusions on the results of an odour impact assessment, and this further complicates the matter. The aim of this paper is to shed light on the criticalities involved in the assessment of the odour impacts from waste management activities. The paper analyses and discusses the potential influences of the choices made during the preparation of dispersion modelling simulations. This contribution is expected to enrich the knowledge base on odour dispersion modelling and to help proponents, environmental consultants and environmental agencies to estimate the impacts induced by current and future waste management operations.

Keywords: odour nuisance, emissions, odour impact, waste treatments, mechanical–biological treatments, dispersion models.

1 INTRODUCTION

Several anthropogenic activities are known causes of environmental pressure, health impacts and nuisance to the settled population. In terms of local impacts, noise and air pollution including particulate matter, are the most evident types of environmental contamination to human beings, who can directly sense those forms of pollution through hearing, the sight and the sense of smell [1]–[6]. The latter allows for the "perception of smell", which is one of the most known definitions of odour [7]. Starting from the last century, several countries worldwide have adopted regulations to limit the ambient air concentrations of various air pollutants, especially macro-pollutants. However, many countries still do not have regulations to limit the impacts of odorants on people, except for prescriptions regarding the comparison of the results of modelling simulations with odour concentration limits during the authorisation procedure of specific activities [8]. Where odour regulations are available, prescriptions are highly variable both among countries and, in some cases, within the same country [9]. In addition, the characteristics of subjectivity of smell perception complicates the matter and differentiates the impacts experienced by the exposed population [10]–[12].

The exposure to malodorous compounds has detrimental effects on human well-being, which are mainly mediated by psychological effects [13]. The prolonged or repetitive exposure to odorants may cause nuisance. The latter, on its turn, may cause a wide variety of adverse effects on human beings, which can be classified as direct or indirect effects: direct

www.witpress.com, ISSN 1743-3541 (on-line)
doi:10.2495/WMEI220101

effects may include psychological symptoms, like irritability, depression, mental confusion [14], and physiological symptoms like headache, sleepiness, nausea, respiratory and cardiovascular effects [15]–[17]; indirect effects derive from direct effects and may lead to negative socio-economic consequences in the geographical context where odour nuisance is experienced. For example, odour nuisance may cause complaints by the exposed population, which can cause the shut-down of the activities that are responsible for the emission of odorants in an area [18]. In tourist areas, odour nuisance may cause the decline of tourism, decrease the quality of life and, in general, odour nuisance may cause the loss of values of houses or terrains and influence the property market [19]–[22].

The problem of odours is related to the exposure to odorant compounds, i.e. substances whose ambient air concentrations exceed their respective odour thresholds in the proximity of target individual (receptor). The compounds characterised by the lowest odour thresholds are typically volatile organic compounds, sulphides, mercaptans, amines, chlorine compounds, indole and skatole [23]. The relationship between the odour concentration of a single compound and its mass concentration in ambient air is generally linear at high mass concentration values (> 1 ppm). However, at lower concentrations, the odour concentration of some pollutants shows a logarithmic trend [24]. This suggests that the overall odour concentration of a mixture may not be predicted by the mass concentrations of the single compounds in the mixture and the related odour threshold values.

While the exposure to odorants is usually associated with reversible symptoms, like those previously reported, the presence of specific compounds in the odorant mixture with adverse effects on humans may cause negative implications on health. Their effects depend on the type of exposure (short-term or long-term), the toxicological characteristics of each compound and the exposure concentration. Long-term exposure to relatively low concentrations should be considered a greater matter of concern if the exposure concentrations are lower than the odour thresholds of the toxic compounds inhaled. This may be the case of benzene, for instance, whose odour threshold determined by Yokio and Nagata [23] is 2.7 ppm. According to the inhalation unit risk provided by the U.S. Environmental Protection Agency [25], the exposure to benzene at a concentration that equals the benzene odour threshold would induce a cancer risk of 7.2×10^{-2}, which exceeds the acceptable cancer risk for exposure to single carcinogenic compounds (1×10^{-6}) by almost five orders of magnitude. In situations where highly toxic compounds are involved, the exposure to odorants may become a matter of human health.

Odour nuisance is caused by various human activities, but it has been especially related to waste management, wastewater treatments and zootechnical activities [10], [26]–[28], which are known sources of the odorant compounds mentioned above. There are many waste processes that cause the problem of odour nuisance, especially waste collection and transportation, waste discharge in dedicated facilities, mechanical–biological treatments (MBTs) of municipal solid waste and landfill disposal. When estimating the impact of odour emissions, the key question is how to properly model each specific emission source. However, there are other crucial choices that may influence the final results and lead to underestimate or overestimate the impacts of specific activities on the environment and the population, e.g. the choice of meteorological data, the meteorological stations, the computational domain and its resolution, the sensitive receptors, whether or not considering terrain elevation, etc.

The present paper moves from the considerations expressed above to discuss the potential influence of input data and modelling choices on the results of the application of modelling techniques for odour dispersion and impact assessment, with a specific focus on odour emissions from waste treatments. After an initial review of relevant publications on the topic of odour dispersion modelling from waste treatment activities, the paper will analyse and discuss the main results of the articles considered and provide indications that may turn useful for further applications of dispersion modelling techniques in this field. Such indications will consider the provisions included in the guidelines on odour impact assessment developed in many countries worldwide. The final aim is two-fold: (1) to help the proponents (and their consultants) of new waste management activities estimate the impacts induced by future waste management operations; and (2) to help the environmental agencies assess the projects according to the environmental impact assessment procedures.

2 GENERALITIES ON DISPERSION MODELLING AND EXAMPLES OF APPLICATIONS TO THE WASTE SECTOR

Odour impacts are normally assessed comparing the results of the application of dispersion models with criteria that depend on the regulations in force in a specific country/region. Dispersion models combine meteorological and morphological input data with the characteristics of the emission sources considered. By implementing equations that describe the mechanisms of convection and diffusion of gases and particles, dispersion models predict the ambient air concentration and the atmospheric deposition to soil of different air pollutants. In the last decades, the technology advances in terms of computational capability have made dispersion modelling affordable to end users at a large scale, simplifying and speeding up the assessment of new or existing activities, while keeping the accuracy of predictions at high levels. This has made dispersion modelling a key step in the environmental impact assessment process [8].

Different types of dispersion models exist, which can be classified into three main categories: gaussian plume, Lagrangian and Eulerian models. Gaussian plume models assume that the distribution of pollutants follows a normal probability distribution in the vertical and in the horizontal directions. They solve the Gaussian plume distribution equation and are particularly used to simulate the dispersion of continuous plumes of pollutants and (in the case of Gaussian puff models) discontinuous sources. Since Gaussian models solve only one equation, they are characterised by fast calculation time. As a drawback, such models must be used with caution when calm wind conditions are frequent and only diffusion occurs [29].

Lagrangian models consider the pollutant as particles that move following buoyancy, wind field and a random process simulating turbulence. Such models calculate the trajectories of a large number of particles by solving ordinary differential equations and use a reference system that follows the particles. Lagrangian models allow for fast calculation time especially in small-scale applications (tens of km), but the computational time increases significantly with the size of the domain and, thus, with the number of trajectories that must be calculated. Puff models area a particular case of Lagrangian and Gaussian models: they simulate the plume as a series of particle releases (puffs), which are treated initially with a Gaussian scheme and then move following a Lagrangian scheme.

Contrarily to Lagrangian models, Eulerian models consider a fixed reference system and the pollutant dispersion is calculated by solving the transport equation. Due to their high computational cost, they are generally used in mesoscale applications to simulate the long-range transport of pollutants and their chemistry [30]. Table 1 presents the most known and used dispersion models according to the approach followed for the computation of air pollutant dispersion.

Originally intended for the prediction of the impact of air pollutants, dispersion modelling have been lately used also to assess the potential impact of odour emissions, as demonstrated

Table 1: List of the most known atmospheric dispersion models used in air quality assessment applications.

Model name	Model type	Developer	License
AERMOD	Gaussian	U.S. Environmental Protection Agency	Open source
ADMS	Gaussian	Cambridge Environmental Research Consultants	Commercial
AUSTAL2000	Lagrangian	Ingenieurbüro Janicke	Open source
CALPUFF	Lagrangian	U.S. Environmental Protection Agency	Open source
CALINE3	Gaussian	U.S. Environmental Protection Agency	Open source
CMAQ	Eulerian	U.S. Environmental Protection Agency	Open source
GEOS-Chem	Eulerian	Harvard and Dalhousie Universities	Open source
HYSPLIT	Lagrangian	National Oceanic and Atmospheric Administration	Open source
ISC3	Gaussian	U.S. Environmental Protection Agency	Open source
WRF-Chem	Eulerian	National Center for Atmospheric Research	Open source

by the large number of publications available in the literature. However, concerning waste management, most papers deal with the wastewater sector. The interest in this sector is visible, for instance, by the recent introduction of the guidelines on odour impacts published by the Lombardy region (Italy), which dedicate a specific document on wastewater treatment plants and propose odour emission factors for specific wastewater treatments [31]. The literature on the application of dispersion models to solid waste treatments and related activities is limited to a small number of publications, mainly concerning the odour impacts of landfills and MBTs, with a minor number of publications on waste collection. In the following sections, recent available publications on odour dispersion modelling applied to waste management activities are reviewed, based on their respective field of application.

The waste management activities are known to contribute to possible odour impacts due to their continuous operation over time. This is the case, for instance, of municipal solid waste (MSW) landfills and MBTs. As a matter of fact, waste collection and transportation are occasional operations that may not affect residents continuatively. In addition, to the authors' knowledge, the application of dispersion modelling to this kind of activities is very limited and the literature lacks studies on this topic. An attempt to define the potential impact of waste transportation was made by Xu et al. [32] who, however, did not consider odour dispersion but only the dispersion of the only ethanol, considered as a representative odorous compound.

One of the problems in modelling landfills is related to the variations in shape and in the surface occupied by active landfills over the simulation period, which, in the case of odour impact and health risk assessments, is normally one year. Another matter of concern is the expected spatial and temporal variability of the odour emission rate, since this may vary depending on the type and age of waste. In their recent work, Szałata et al. [33] solved these issues by implementing a trial-and-error procedure with CALPUFF to estimate the specific odour emission rate (SOER) adopted in their case study, which resulted as 5.1 $ou_E/m^2/s$ (as a matter of fact, landfill sites are normally modelled as area sources). The trial-and-error procedure was based on odour sampling and following determination of the odour concentration of the sample by dynamic olfactometry (EN 13725:2003) in the vicinity of the landfill site. Dynamic olfactometry is a technique that considers a panel of selected examiners, who are subjected to increasing concentrations of the sample, which is diluted with fresh air by an olfactometer. The odour concentration (C_{od}) of the sample corresponds to the geometric mean of the number of dilutions that are necessary to each panellist to perceive the odour [34].

Alternatively, a more direct approach can be adopted [35]: a flux chamber and Nalophan® sampling bags can be used to take odour samples directly on representative portions of the landfill surface, according to the EN 13725:2003 standard. The samples can be analysed by dynamic olfactometry to determine the C_{od} of each sample, expressed as European odour units (ou_E) per m^3. Based on the inlet flow rate of the flux chamber (Q, m^3/s) and its sampling area (A_W, m^2), the SOER ($ou_E/m^2/s$) can be calculated as follows:

$$SOER = \frac{Q \cdot C_{od}}{A_W}. \tag{1}$$

In a recent work, Tagliaferri et al. [36] analyse how specific choices can affect the results of dispersion models applied to waste landfills. According to the authors, the most critical aspect is the characterisation of the emission sources, in terms of source geometry and odour emission rate (OER). Regarding source geometry, dispersion models require the definition of an initial vertical dispersion coefficient ($\sigma_{z,0}$, expressed as m), representing the initial vertical dimension of the plume. In the case of area sources (e.g., landfills), the choice of this value is particularly difficult: landfills are characterised by large surface areas, above which different degrees of turbulence may occur. In the work by Tagliaferri et al. [36], different choices of $\sigma_{z,0}$ generated odour concentrations at the receptor points that differed by a factor of three.

In a recent paper, Toledo et al. [37] evaluated the dispersion of odour from waste composting. The authors assume the plant as a point source with an OER calculated multiplying the SOER (obtained through odour sampling on the waste piles and following olfactometric analysis) by the emission area of waste piles. In another paper [38] on an anaerobic digestion plant for municipal solid waste, the following emission sources were identified: a heap of yard waste and a biofilter for air pollution and odour control. Both the sources were modelled as area sources; the biofilter was characterised by an output velocity of the treated effluent, while free convection was assumed for the yard waste heap. SOER were estimated by applying a flux chamber on the waste yard heap and on the biofilter surface. The authors compared the results of the application of CALPUFF on the anaerobic digestion plant with field measurements and found good agreement between the two solutions.

When present, biofilters are one of the main sources of odour in MBT plants. They are intended as a pollution control technology for polluted air streams that can be conveyed. For this reason, they have been largely employed in aerobic MBTs of waste, where large airflow rates are needed to stabilise the waste biomass.

Luciano et al. [39] applied CALPUFF to simulate the dispersion of odour from two biofilters to investigate the effects of the enlargement of an existing MBT plant. The authors derived the SOER through odour sampling and dynamic olfactometry in one case (EN 13725: 2003) and applying the local concentration limit value of 300 ou_E/m^3 on the biofilter surface in another case [31]. A different model (AERMOD) was used in a publication on municipal solid waste bio-drying [40]. In this publication, a conventional open-bed biofilter was compared with an array of bio-trickling filters to evaluate the potential improvements in the surroundings of a plant In terms of air quality and odour impact. The biofilter was simulated as an active area source, while the bio-trickling filters were simulated as point sources. Indeed, bio-trickling filters can convey the treated air stream and release it from a stack. The advantages of bio-trickling filters are noteworthy: the higher outlet velocity of the treated air released by bio-trickling filters guarantees a stronger dispersion of the pollutants and odorants in the atmosphere than biofilters, whose outlet velocity is about 200 times lower. The simulations revealed that a replacement of an open biofilter with an equivalent array of bio-trickling filters may reduce the impacts in the vicinity of the plant by > 90%.

3 DISCUSSION AND BEST PRACTICES FOR FUTURE APPLICATIONS

The experiences that are described in the papers presented in the previous section allow elaborating a discussion for future applications of dispersion modelling to waste-related activities. Part of the following advice is also included in the guidelines and regulations on odour impact assessment of various countries worldwide [9].

3.1 Computational domain

Defining a computational domain for a simulation means selecting its size, the horizontal and vertical resolution and the location of sensitive receptors. The extension of the domain should be chosen to include the areas where the highest impact is expected, and all the sensitive receptors identified. Depending on the model, the user can choose between a cartesian or a polar grid. However, some guidelines/regulations on odour impact make explicit reference to cartesian grids. As a rule, the horizontal grid cell size should be always lower than the minimum distance between the emission source(s) and the nearest sensitive receptor. Since odour impact is generally evaluated at a local scale, an adequate value of cell size should be comprised between 25 and 250 m [31], [41]. However, it is preferable to determine the horizontal cell size through a sensitivity analysis, reducing the cell size until the location of the areas with the highest impacts become independent of the cell size itself.

The definition of sensitive receptors allows for the calculation of statistics on the C_{od} expected in the selected points. This step is essential to verify the compliance with the regulations in force. Different guidelines/regulations provide instructions on the definition of receptor points: for instance, guidelines/regulations may specify the kinds of receptors (e.g., the nearest house, schools, hospitals), the minimum number and the distribution of receptors around the emission source(s) to be considered.

3.2 Meteorological data

Air dispersion models require the pre-processing of meteorological data to compute the wind field for every timestep. The nature of the meteorological data depends on the dispersion model chosen. Normally, dispersion models give two options: the possibility of using a wind field library generated by an external meteorological model (e.g., the weather research and forecasting model (WRF)) and the possibility of generating a wind field library with the model's meteorological pre-processor. The second option requires the availability of surface meteorological observations (recorded by meteorological or air quality stations) near the emission sources to be modelled.

Some guidelines/regulations on odour impact assessment set specific requirements on the maximum distance of meteorological stations from the nearest emission source and on the locations of the meteorological stations that can be considered as representative of the area. For instance, in Italy, regional guidelines require at least one meteorological station at a distance < 10 km from the nearest emission source. In the case of complex orography [41], [42], the meteorological station(s) must be located in the same valley of the emission sources [31], [43]. The same guidelines set minimum requirements on data quality (e.g., minimum anemometer elevation, minimum frequency of observations, minimum simulation period, maximum percentage of invalid data). In addition to surface observations, some dispersion models, like CALPUFF or AERMOD, require upper air soundings. In case the aforementioned requirements cannot be fulfilled, meteorological data should be retrieved by applying an external meteorological model.

3.3 Elevation data

Orography significantly affects the results of a dispersion model. The presence of terrain elevation is responsible for altering the wind field compared to a situation with flat terrain. According to the guidelines of the Lombardy region (Italy), for instance, terrain elevation should be taken into account if the ratio between the maximum difference in elevation within the computational domain and the size of the minimum horizontal dimension of the domain exceeds the value of 0.01 [31]. The Shuttle Radar Topography Mission (SRTM) of the U.S. Geological Survey (USGS) has made elevation data available for many countries worldwide [44]. The data are available for download upon registration.

The presence of buildings can also be taken into account in most of the dispersion models available. Like orography, buildings may affect the dispersion of air pollutants and odorants. The phenomenon of building downwash occurs when an emission source is located upwind of a building, whose height and/or width are sufficiently high to generate a turbulent flow downstream that can bring the plume to the ground and increase the impacts at ground level [45]. To account for this phenomenon, many available dispersion models contain a module for building downwash calculation that requires the explicit definition of the buildings near the emission source. However, the presence of buildings may also affect areas that are far from the emission sources, since it modifies the surface roughness of the ground. For this reason, many dispersion models allow defining the spatial distribution of surface roughness in an explicit way or requiring land use maps as input files. Peculiar situations (e.g., emission of air pollutants or odour in urban areas) may require the explicit definition of buildings for a proper evaluation of the population exposure. This is particularly useful when assessing the impacts of transportation in canyon-like streets [46]. Among the most known dispersion models, AUSTAL2000 allows the user to create and import a raster map of building elevations to explicitly define buildings within a portion of the computational domain.

3.4 Emission characterisation

The complete characterisation of emission sources is generally challenging and is usually affected by high uncertainties, especially in the case of diffusive sources. The definition of conveyed sources is an easier task: for this kind of sources, OER values can be measured by odour sampling at the source and dynamic olfactometry; alternatively, odour emission factors (or C_{od} values to be multiplied by the airflow rate) can be retrieved in the literature or in previous odour impact assessment studies, if similar cases have been studied. However, if the purpose is to assess the maximum permissible impacts within the computational domain, the user should consider the OER limit values in force in the country/region. On the contrary, diffusive or passive sources are normally unpredictable and, in order to obtain the best results, the definition of OER (or SOER for area sources) values requires the measurement of odour concentrations.

In the case of waste landfills, due to the expected spatial and temporal variability of the odour emission rate from the landfill surface, olfactometric analyses play a fundamental role in the definition of a representative SOER. The latter could be determined directly by sampling the landfill surface via wind tunnels or flux chambers. According to Lucernoni et al. [47], flux chambers are preferable, due to the tendency of wind tunnels to overestimate the emission rate. The final SOER value could be defined as the geometrical mean of the SOER values measured in different points of the landfill surface [36]. If the landfill surface is not accessible, the SOER can be determined indirectly, by following the procedure adopted by Szałata et al. [33]: assuming tentative values of SOER in the model and selecting the value that allows the model results to match the results of the olfactometric analyses carried out in the vicinity of the landfill site.

Model validation, performed by on-field sampling and dynamic olfactometry, plays a key role in the choice of the best parameters that characterise the emission source, including $\sigma_{z,0}$ [36]. Brancher et al. [48] stress the importance of considering hourly-resolved OER values when evaluating low percentiles of odour concentration at receptor points. By implementing a stochastic approach based on a Monte Carlo method, the authors proved this choice to deliver more robust results. However, when high-end percentiles are the terms of reference of odour impact criteria, assuming a constant OER is preferable, since the effort to generate hourly-resolved OER values is not adequately paid back by the quality of the results.

Besides diffusive sources that can be logically modelled as area sources (e.g., landfills or waste heaps in open air), there are other sources that require particular attention and, possibly, a different approach. This is the case, for instance, of warehouses or buildings containing waste, with permanent lateral openings or, alternatively, portals that are normally kept closed, but that can be opened during the operation of a waste treatment facility. Inside these buildings, odorants are released by the waste that is stored or moved by machineries. Such buildings become emission sources when the portals are opened or if lateral openings are present. The characterisation of such sources is challenging and the literature lacks publications on this topic. According to the user's guides of the most known dispersion models, this kind of sources could be modelled as volumetric sources, i.e. sources for which it is possible to define both initial vertical ($\sigma_{z,0}$) and horizontal ($\sigma_{y,0}$) dispersion coefficients. In the case of odour emissions from a building, useful indications on how to define volumetric sources are provided by the Ministry for the Environment of New Zealand [49]: as a first approximation, $\sigma_{z,0}$ and $\sigma_{y,0}$ can be assumed respectively as the building height and the building minimum lateral dimension, both multiplied by 0.25.

The definition of the OER is also a critical point when dealing with volumetric sources. Contrarily to point or area sources, for which OER may be obtained by similar cases in the literature or by odour sampling and dynamic olfactometry, situations that may be modelled as volumetric sources are less common and difficult to generalise. However, in many practical odour impact assessment procedures, the problem of the definition of the OER is usually treated as follows:

- Firstly, it is necessary to retrieve a representative C_{od} value in the indoor environment. Odour sampling and dynamic olfactometry, though more time consuming and expensive, are expected to deliver the best estimation. Alternatively, this can be done by retrieving C_{od} values related to the same kind of waste in the literature or in similar case studies.
- Secondly, it is necessary to estimate the air exchange rate between indoors and outdoors. This this step is particularly affected by uncertainties. An empirical approach consists of considering the number of times the building door is opened during the day, the vertical area of the building opening, the mean wind speed at the nearest anemometer or grid point where meteorological data were calculated by the meteorological pre-processor. This way it is possible to estimate the volumetric airflow rate through the opening, which can be assumed as equal to the volumetric airflow rate of the air exiting the building. As

a precautionary measure, in a first approximation, the wind direction might be assumed as perpendicular to the building opening.

- Finally, the OER (ouE/s) can be estimated multiplying C_{od} by the volumetric airflow rate calculated at the previous step. Volume sources may be used also to simulate the movement of vehicles if the dispersion model does not allow defining line sources. In this case, it is suggested to define a series of multiple volumetric sources separated by < 0.25 times the distance to the nearest receptor point.

4 CONCLUSIONS

The present paper discussed the role of dispersion modelling for odour impact assessment applied to waste management activities. The paper highlighted some critical issues in the application of dispersion models. Despite the advances achieved over the last decades in terms of programming and computational capability, modelling the phenomena of odour emission, dispersion and perception by humans is still challenging and is affected by a high degree of uncertainty. This is due in part to the subjectivity of human perception and in part to the chemical composition of odour emissions. The latter may change along the path of the plume. In addition, the odour concentration of some compounds was found not to increase linearly with their mass concentration in air. There is also a great uncertainty in the characterisation of emission sources, especially diffusive sources that are normally treated as passive area and volumetric sources. The measurement of odour concentration and OER by sampling and dynamic olfactometry still plays a key role in the validation of model results and should not be replaced entirely by dispersion modelling, except for the case when the same dispersion model is applied more times to the same site where it was previously validated. Although the suggestions and recommendations here discussed are proposed for application to waste management activities and odour impact assessment, they may be equally applied to any other field and pollutants.

REFERENCES

[1] Adamiec, E., Jarosz-Krzemińska, E. & Bilkiewicz-Kubarek, A., Adverse health and environmental outcomes of cycling in heavily polluted urban environments. *Scientific Reports*, **12**(1), p. 148, 2022.

[2] Paolocci, G., Bauleo, L., Folletti, I., Murgia, N., Muzi, G. & Ancona, C., Industrial air pollution and respiratory health status among residents in an industrial area in central Italy. *International Journal of Environmental Research and Public Health*, **17**(11), p. 3795, 2020.

[3] Istrate, I.A., Oprea, T., Rada, E.C. & Torretta, V., Noise and air pollution from urban traffic. *WIT Transactions on Ecology and the Environment,* vol. 191, WIT Press: Southampton and Boston, pp. 1381–1389, 2014.

[4] Torretta, V., Raboni, M., Copelli, S., Rada, E.C., Ragazzi, M., Ionescu, G., Apostol, T. & Badea, A., Application of strategies for particulate matter reduction in urban areas: An Italian case. *UPB Scientific Bulletin, Series D: Mechanical Engineering,* **75**(4), pp. 221–228, 2013.

[5] Ciuta, S., Schiavon, M., Chistè, A., Ragazzi, M., Rada, E.C., Tubino, M., Badea, A. & Apostol, T., Role of feedstock transport in the balance of primary PM emissions in two case-studies: RMSW incineration vs. sintering plant. *UPB Scientific Bulletin, Series D: Mechanical Engineering*, **74**(1), 2pp. 11–218, 2012.

[6] Schreckenberg, D., Griefahn, B. & Meis, M., The associations between noise sensitivity, reported physical and mental health, perceived environmental quality, and noise annoyance. *Noise and Health*, **12**(46), pp. 7–16, 2010.

[7] Conti, C., Guarino, M. & Bacenetti, J., Measurements techniques and models to assess odor annoyance: A review. *Environment International*, **134**, 105261, 2020.

[8] Invernizzi, M., Brancher, M., Sironi, S., Capelli, L., Piringer, M. & Schauberger, G., Odour impact assessment by considering short-term ambient concentrations: A multi-model and two-site comparison. *Environment International*, **144**, 105990, 2020.

[9] Bokowa, A., Diaz, C., Koziel, J.A., McGinley, M., Barclay, J., Schauberger, G., Guillot, J.-M., Sneath, R., Capelli, L., Zorich, V., Izquierdo, C., Bilsen, I., Romain, A.C., del Carmen Cabeza, M., Liu, D., Both, R., Van Belois, H., Higuchi, T. & Wahe, L., Summary and overview of the odour regulations worldwide. *Atmosphere*, **12**, p. 206, 2021.

[10] Piccardo, M.T., Geretto, M., Pulliero, A. & Izzotti, A., Odor emissions: A public health concern for health risk perception. *Environmental Research*, **204**, 112121, 2022.

[11] Li, H., Qi, G., Liu, X., Ren, L., Zhao, Y. & Sun, Y., Emission characteristics and health risk assessment of odorous pollutants from organic fraction of municipal solid waste compost in summer. *Research of Environmental Sciences*, **33**(4), pp. 868–875, 2020.

[12] Tjalvin, G., Magerøy, N., Bråtveit, M., Lygre, S.H.L., Hollund, B.E. & Moen, B., Odour as a determinant of persistent symptoms after a chemical explosion, a longitudinal study. *Industrial Health*, **55**(2), pp. 127–137, 2017.

[13] Zarra, T., Belgiorno, V. & Naddeo, V., Environmental odour nuisance assessment in urbanized area: Analysis and comparison of different and integrated approaches. *Atmosphere*, **12**(6), p. 690, 2021.

[14] Schiffman, S.S., Miller, E.A.S., Suggs, M.S. & Graham, B.G., The effect of environmental odors emanating from commercial swine operations on the mood of nearby residents. *Brain Research Bulletin*, **37**(4), pp. 369–375, 1995.

[15] Jacquemin, B., Sunyer, J., Forsberg, B., Götschi, T., Bayer-Oglesby, L., Ackermann-Liebrich, U., de Marco, R., Heinrich, J., Jarvis, D., Torén, K. & Künzli, N., Annoyance due to air pollution in Europe. *International Journal of Epidemiology*, **36**(4), pp. 809–820, 2007.

[16] Aatamila, M., Verkasalo, P.K., Korhonen, M.J., Suominen, A.L., Hirvonen, M.-R., Viluksela, M.K. & Nevalainen, A., Odour annoyance and physical symptoms among residents living near waste treatment centres. *Environmental Research*, **111**(1), pp. 164–170, 2011.

[17] Claeson, A.-S., Lidén, E., Nordin, M. & Nordin, S., The role of perceived pollution and health risk perception in annoyance and health symptoms: A population-based study of odorous air pollution. *International Archives of Occupational and Environmental Health*, **86**(3), pp. 367–374, 2013.

[18] Keck, M., Mager, K., Weber, K., Keller, M., Frei, M., Steiner, B. & Schrade, S., Odour impact from farms with animal husbandry and biogas facilities. *Science of the Total Environment*, **645**, pp. 1432–1443, 2018.

[19] Wojnarowska, M., Sagan, A., Plichta, J., Plichta, G., Szakiel, J., Turek, P. & Sołtysik, M., The influence of the methods of measuring odours nuisance on the quality of life. *Environmental Impact Assessment Review*, **86**, 106491, 2021.

[20] Errajaa, K., Legohérel, P., Daucé, B. & Bilgihan, A., Scent marketing: linking the scent congruence with brand image. *International Journal of Contemporary Hospitality Management*, **33**(2), pp. 402–427, 2021.

[21] Wojnarowska, M., Sołtysik, M., Sagan, A., Stobiecka, J., Plichta, J. & Plichta, G., Impact of odor nuisance on preferred place of residence. *Sustainability*, **12**(8), p. 3181, 2020.

www.witpress.com, ISSN 1743-3541 (on-line)

[22] Beloff, B.R., Beaver, E.R. & Massin, H., Assessing societal costs associated with environmental impacts. *Environmental Quality Management,* **10**(2), pp. 67–82, 2000.
[23] Yokio, Y. & Nagata, E., Measurement of odor threshold by triangle odor bag method. https://www.env.go.jp/en/air/odor/measure/02_3_2.pdf. Accessed on: 26 Jan. 2022.
[24] Schiavon, M., Martini, L.M., Corrà, C., Scapinello, M., Coller, G., Tosi, P. & Ragazzi, M., Characterisation of volatile organic compounds (VOCs) released by the composting of different waste matrices. *Environmental Pollution*, **231**, pp. 845–853, 2017.
[25] United States Environmental Protection Agency, Benzene – CASRN 71-43-2, DTXSID3039242. https://cfpub.epa.gov/ncea/iris2/chemicallanding.cfm?substance_nmbr=276. Accessed on: 12 Mar. 2022.
[26] Rappert, S. & Müller, R., Odor compounds in waste gas emissions from agricultural operations and food industries. *Waste Management*, **25**(9), pp. 887–907, 2005.
[27] Liu, Y., Chen, J., Yang, H., Wang, J. & Zou, K., Emission of pollutants and odor pollution at initial decomposition stage of municipal solid waste. *Research of Environmental Sciences*, **35**(1), pp. 238–245, 2022.
[28] Senatore, V., Zarra, T., Galang, M.G., Oliv, G., Buonerba, A., Li, C.W., Belgiorno, V. & Naddeo, V., Full-scale odor abatement technologies in wastewater treatment plants (WWTPs): A review. *Water*, **13**(24), p. 3503, 2021.
[29] Leelőssy, Á., Molnár, F., Izsák, F., Havasi, Á., Lagzi, I. & Mészáros, R., Dispersion modeling of air pollutants in the atmosphere: A review. *Central European Journal of Geosciences*, **6**(3), pp. 257–278, 2014.
[30] Li, X.-X., Liu, C.-H., Leung, D.Y.C. & Lam, K.M., Recent progress in CFD modelling of wind field and pollutant transport in street canyons. *Atmospheric Environment*, **40**, pp. 5640–5658, 2006.
[31] Regione Lombardia, Linea guida per la caratterizzazione, l'analisi e l'autorizzazione delle emissioni gassose in atmosfera delle attività ad impatto odorigeno. https://www.regione.lombardia.it/wps/portal/istituzionale/HP/DettaglioServizio/servizi-e-informazioni/Imprese/Sicurezza-ambientale-e-alimentare/qualita-dell-aria-ed-emissioni-in-atmosfera/emissioni-odorigene-linee-guida-di-settore/emissioni-odorigene-linee-guida-di-settore. Accessed on: 24 Feb. 2022.
[32] Xu, A., Chang, H., Zhao, Y., Tan, H., Wang, Y., Zhang, Y., Lu, W. & Wang, H., Dispersion simulation of odorous compounds from waste collection vehicles: Mobile point source simulation with ModOdor. *Science of the Total Environment*, **711**, 135109, 2020.
[33] Szałata, Ł., Zwoździak, J., Majerník, M., Cierniak-Emerych, A., Jarossová, M.A., Dziuba, S., Knošková, L.'. & Drábik, P., Assessment of the odour quality of the air surrounding a landfill site: A case study. *Sustainability*, **13**, p. 1713, 2021.
[34] Sironi, S., Capelli, L., Céntola, S., Del Rosso, R. & Pierucci, S., Odour impact assessment by means of dynamic olfactometry, dispersion modelling and social participation. *Atmospheric Environment*, **44**, pp. 354–360, 2010.
[35] Naddeo, V., Zarra, T., Oliva, G., Chiavola, A. & Vivarelli, A., Environmental odour impact assessment of landfill expansion scenarios: Case study of Borgo Montello (Italy). *Chemical Engineering Transactions*, **54**, pp. 73–78, 2016.
[36] Tagliaferri, F., Invernizzi, M., Sironi, S. & Capelli, L., Influence of modelling choices on the results of landfill odour dispersion. *Detritus*, **12**, pp. 92–99, 2020.
[37] Toledo, M., Gutiérrez, M.C., Siles, J.A. & Martín, M.A., Full-scale composting of sewage sludge and market waste: Stability monitoring and odor dispersion modelling. *Environmental Research*, **167**, pp. 739–750, 2018.

[38] Ranzato, L., Barausse, A., Mantovani, A., Pittarello, A., Benzo, M. & Palmeri, L., A comparison of methods for the assessment of odor impacts on air quality: Field inspection (VDI 3940) and the air dispersion model CALPUFF. *Atmospheric Environment*, **61**, pp. 570–579, 2012.

[39] Luciano, A., Torretta, V., Mancini, G., Eleuteri, A., Raboni, M. & Viotti, P., The modelling of odour dispersion as a support tool for the improvements of high odours impact plants. *Environmental Technology*, **38**(5), pp. 588–597, 2017.

[40] Schiavon, M., Ragazzi, M., Torretta, V. & Rada, E.C., Comparison between conventional biofilters and bio-trickling filters applied to waste bio-drying in terms of atmospheric dispersion and air quality. *Environmental Technology*, **37**(8), pp. 975–982, 2016.

[41] Adami, L., Schiavon, M. & Rada, E.C., Potential environmental benefits of direct electric heating powered by waste-to-energy processes as a replacement of solid-fuel combustion in semi-rural and remote areas. *Science of the Total Environment*, **740**, 140078, 2020.

[42] Schiavon, M., Adami, L., Torretta, V. & Tubino, M., Environmental balance of an innovative waste-to-energy plant: The role of secondary emissions. *International Journal of Environmental Impacts*, **3**(1), pp. 84–93, 2020.

[43] Regione Veneto, Indicazioni per l'utilizzo di tecniche modellistiche per la simulazione della dispersione di inquinanti in atmosfera. https://www.arpa.veneto.it/temi-ambientali/aria/file-e-allegati/applicazioni-modellistiche/Indicazioni_tecniche_ modellistiche_simulazioni_atmosfera.pdf. Accessed on: 2 Mar. 2022.

[44] EarthExplorer, U.S. Geological Survey. https://earthexplorer.usgs.gov/. Accessed on: 2 Mar. 2022.

[45] Yang, B., Gu, J. & Zhang, K.M., Parameterization of the building downwash and sidewash effect using a mixture model. *Building and Environment*, **172**, 106694, 2020.

[46] Schiavon, M., Redivo, M., Antonacci, G., Rada, E.C., Ragazzi, M., Zardi, D. & Giovannini, L., Assessing the air quality impact of nitrogen oxides and benzene from road traffic and domestic heating and the associated cancer risk in an urban area of Verona (Italy). *Atmospheric Environment*, **120**, pp. 234–243, 2015.

[47] Lucernoni, F., Tapparo, F., Capelli, L. & Sironi, S., Evaluation of an odour emission factor (OEF) to estimate odour emissions from landfill surfaces. *Atmospheric Environment*, **144**, pp. 87–99, 2016.

[48] Brancher, M., Hoinaski, L., Piringer, M., Prata, A.A. & Schauberger, G., Dispersion modelling of environmental odours using hourly-resolved emission scenarios: Implications for impact assessments. *Atmospheric Environment: X*, **12**, 100124, 2021.

[49] Ministry for the Environment of New Zealand, Good practice guide for atmospheric dispersion modelling. http://tools.envirolink.govt.nz/assets/Uploads/Good-Practice-Guide-MFE-atmospheric-dispersion-modelling-jun04.pdf. Accessed on: 1 Mar. 2022.

SECTION 4
WASTE MONITORING AND MANAGEMENT

PROPOSAL FOR THE LOCATION OF A WASTE MANAGEMENT CENTER ON A UNIVERSITY CAMPUS: A CASE STUDY

BETHY MERCHÁN-SANMARTIN[1,2], ERICK GUADAMUD-SOLEDISPA[1,2],
GIANELLA BRAVO-MURILLO[1,2] & PAÚL CARRIÓN-MERO[1,2]
[1]Center for Research and Projects Applied to Earth Sciences (CIPAT), ESPOL Polytechnic University, Ecuador
[2]Faculty of Engineering in Earth Sciences (FICT), ESPOL Polytechnic University, Ecuador
[3]Faculty of Engineering Sciences, Santa Elena Peninsula State University (UPSE), Ecuador

ABSTRACT

Garbage management and the use of waste are part of sustainable development goals (SDGs) 8, 11, and 12. The recycling process is simple and contributes to improving the environment. Therefore, garbage management in Ecuador is a relevant issue that society must address regardless of the size of the community that generates it. In this case, university campuses need services and infrastructure, including waste management at the scale of a small city, because they require services like these (e.g., accommodation, transportation, cleaning, and waste management). This research seeks to propose the ideal location for a separation, assessment, and recycling center (CSVR), through the implementation of a multi-criteria matrix, with the application of the vertical electrical sounding techniques (SEVs), for the sustainable management of waste. The methodological process includes: (i) identification of probable sites; (ii) application of a multi-criteria matrix for CSVR; (iii) application of geophysical methods (SEVs); and (iv) interpretation of results and selection of suitable areas for the recycling center. Six sites were analyzed, of which three were ideal for the implementation of SEVs where there will be control and management of invasions, as well as non-contamination of water resources. All the selected areas are permeable in their superficial layer and favor leaching; therefore, waterproofing at its base and some design criteria that contemplate technical and environmental monitoring were recommended.

Keywords: garbage, recycling, water resources, vertical electrical surveys, resource recovery.

1 INTRODUCTION

Garbage or rubbish generation is a problem that affects the entire society due to its exponential growth [1]. This growth is mainly conditioned by per capita income, weather conditions, geographic areas, and cultural issues. In general, garbage is classified into two large groups: (i) trash, which includes that part of the garbage that is no longer susceptible to any use; and (ii) waste or residue, that garbage that can become a resource for a new good and service [2].

The pollution generated by a poor garbage disposal, or inefficient waste and residue management, also grows significantly [3]. This situation is aggravated by the difficult access to collection vehicles in marginal areas; additionally, this means increasing the unit price of waste collection and generating the illegal disposal of household waste that can cause significant problems to the environment and water resources, contaminating them with leachate [4]. First-world countries have considered this problem and have acted to minimize it [5]. In the European Union (EU) countries, suitable containers can be observed to separate and assess each waste, which can be used to manage a circular and sustainable economy [6]. The purpose of good waste and residue management is to seek mainly to reduce the amount produced, then it is about minimizing the amount of unclassified garbage eliminated, to finally increase the recycling of materials [7].The entities in charge manage the management of garbage produced by the community, generally local governments [8]. However, its

WIT Transactions on Ecology and the Environment, Vol 257, © 2022 WIT Press
www.witpress.com, ISSN 1743-3541 (on-line)
doi:10.2495/WMEI220111

resources are limited and are directed to other areas; therefore, collection routes do not increase in depressed and/or marginal areas [9].

The decomposition of garbage generates leachates, which are liquids caused by the presence of organic matter, between 40% and 70% of the garbage [10]. In addition, they can be caused by factors such as the composition of the waste, discharge management, climatic conditions, the physical–chemical characteristics of the landfills, the age of the landfill, and biochemical demands [11]. Often, leachates have extreme pH, high chemical oxygen demand (COD) and biochemical oxygen demand (BOD), high concentration of inorganic salts, and relevant toxicity [12]. It should be remembered that leachate is the primary source of contamination of water resources, mainly underground [13], [14]. Less than 3% of fresh water is available on planet earth; Of this percentage, 30% corresponds to fresh groundwater [15]. One of the ways to determine the presence or not of underground water resources and the composition of the soil is through vertical electrical soundings (SEVs) [16]. SEVs are a practical method with a significant contribution due to their relevance in geological and hydrogeological studies. They can be applied for exploration projects in the depths of the underground layers in predetermined areas. It is necessary to investigate weathered, sedimentary material, rocks, runoff, and soil resistivity [17].

The study of SEVs serves to identify vertical sections in the upper layer of the soil; it generally sends electrical waves that analyze the resistivity variations of the material in the study area [18]. The electrical sounding of the soil is given to various nodes that are scattered within the analysis area, which allows understanding of the behavior of the soil and the type of existing material; all this is due to the device known as a Tellurometer, introducing alternating current using electrodes [19]. The electrical resistivity produced by a Tellurometer in SEVs studies serves to recognize the number of pores present in the different soil layers [20]. Additionally, in selecting a suitable site for the CSVR project, it is necessary to implement a decision matrix. It is a quantitative scheme that helps the team or subject analyze and identify the correct relationships between the set of project information; the evaluation criterion is provided by the analysis of the mode of action and a systematic consistency check [21].

Finally, geographic information systems (GIS) are applied, models with various geo-processing tools in ArcGIS, which generate several tools for geotechnical, environmental, topographic, road, and hydrological studies [22], [23]. The most used tools, according to information sampling, are raster DEM (digital terrain model for numerical data structure, which represents the spatial distribution of a quantitative and continuous variable) and TIN (digital elevation models to represent the morphology of the surface) [24], [25].

The Escuela Superior Politécnica del Litoral (ESPOL) in Ecuador, has a university campus that generates waste and collects it without proper processing [26]. This situation requires that the creation of a CSVR be addressed by running SEVS, applying a decision matrix, and being supported by GIS. In this case, using SEVs for a CSVR helps identify suitable areas far from the aquifers so as not to contaminate them with leachate or non-environmental factors [27]. Worldwide, especially in Latin American and Caribbean countries, the management and use of solid waste have been improving through investments and continuing education on the subject [28]. In neighboring countries, such as Colombia, the design proposal for a comprehensive recycling center, including alternative energy sources, has already been presented [29].

The purpose of this work is to secure the ideal location for a separation, assessment, and recycling center (CSVR), through the implementation of a decision matrix, with the application of the vertical electrical sounding technique (SEVs), which will delimit the hydrogeological zoning and suitable sites that present the best conditions, for the sustainable management of waste and use of residues in the university.

2 LOCATION OF THE STUDY AREA

ESPOL is the number one public university in Ecuador. It is located in the center-north of the city of Guayaquil, located on the Gustavo Galindo Campus, with the topographic zone 17 S on the Universal Transversal Mercator (UTM) scale. It coordinates East: 615230.56 m E, North: 9762623.83 m S (Fig. 1). The Campus has a size of 675.35 ha, where 332.30 ha are from the Prosperina protective forest, and 343.05 ha are the sites that could be intervened. Its topography varies with heights between 35 to 210 m above sea level. From a geological point of view, the Piñón formation (Jurassic) are emissions of volcanic flows with extensions in the oceanic crust, often produced by basaltic lavas "pillow lavas, and various volcanic intercalations" [30], [31]. However, at the local level, according to Sánchez Padilla et al. [32], it establishes that the Cayo Member, represented by volcanic materials, is the one that outcrops in the most significant extension of the study area.

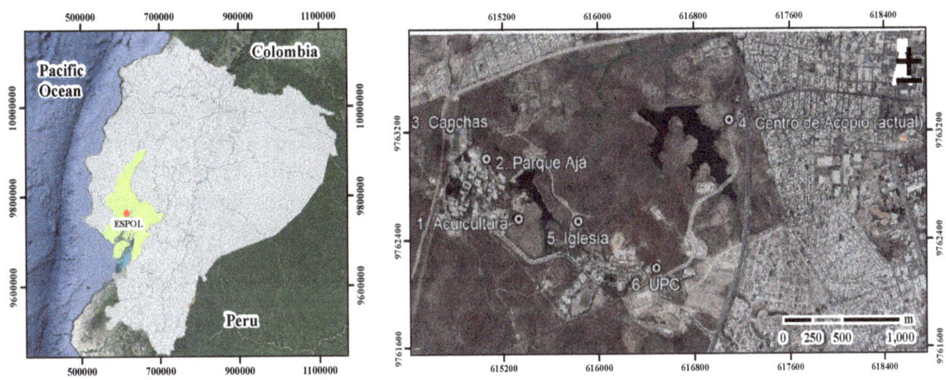

Figure 1: Study area.

3 MATERIALS AND METHODS

The implemented methodology is based on the scheme of Fig. 2, therefore there are the following phases related to the project. In Phase I of identification and information collection, the base data for the geophysical analysis within the Gustavo Galindo campus was obtained. This information includes the search for orthophotos, isohyet planes, hydrological data, soil and rock types, seismicity, and groundwater collection points. These data were obtained from the National Institute of Meteorology and Hydrology (INAMHI); the Ministry of Agriculture, Livestock, Aquaculture and Fisheries (MAGAP). In addition, the Google Earth application was used to obtain updated satellite images of the area. The information and data provided were analyzed with which a morphological map, infiltration analysis and landslide susceptibility map of the site were made.

The realization of the morphological map begins with the map of slopes in percentage. It is essential because it reflects the degree of inclination in analyzing its relief. According to the Agustín Codazzi Geographical Institute, within the Land Classification for its capacity to use, it is considered that 12% of the slope gradient is a critical point for mechanization. From this value, the susceptibility to soil erosion increases.

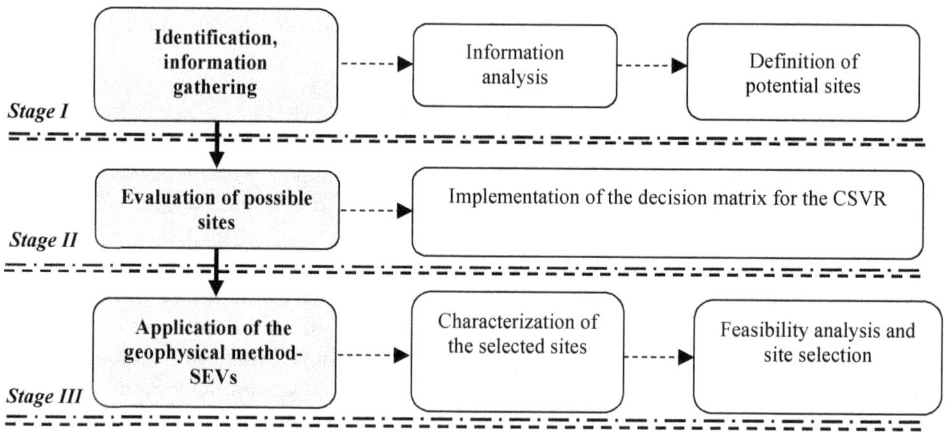

Figure 2: General scheme of implemented methodology.

The orthophotos, the water network, and the DEM were used in the infiltration analysis. The possible zones were created with a 50 m buffer, considering the MAGAP regulations, which suggest this distance for safety to carry out a study in case of any contamination.

In the case of mass susceptibility, also known as landslides, the Mora–Vahrson methodology is used, which is applicable in development zones because it requires few morphodynamic variables, and its application is relatively simple. The most significant factors are included from the point of view of slope instability, which is based on parameters that can be determined in an agile way.

With the results of these three criteria, it was possible to define six probable sites for evaluation in stage II.

In Phase II of evaluating possible sites, the method used uses a decision matrix (MCP), which seeks to facilitate the analysis of problems in a situation that is of great uncertainty or has inaccuracies of the specific terrain when it has a very high degree of complexity. Optimization through MCP must have calculations that involve critical criteria and evaluations that drive the final decisions on the optimal alternatives.

One of the fundamental parts of the segmentation process is determining the influence factors for quantifying results. The first step in creating a good MCP is to have a structure with analysis factors such as population, geotechnical, topographical, water bodies, and route alternatives. Considering the six locations (see Fig. 1), the MCP considers the following:

1. It must be far from runoff and infiltration areas to avoid contamination by leachate.
2. The CSVR must be distanced from the polytechnic campus's central "populated" area.
3. It cannot be considered located within a possible landslide hazard site.
4. The topography should be smooth with no slope.
5. The location must consider the route of the collection truck.

From these five factors, the MCP index is obtained, which allows the possible locations to be identified quantitatively (eqns (1) and (2)).

$$\text{MCP} = (\text{Coef. P.})\text{Population} + (\text{Coef. H.})\text{Hazard} + (\text{Coef. R.})\text{Runoff} + (1)(\text{Coef. T.})\text{Topography} + (\text{Coef. R.})\text{Route}. \quad (1)$$

*The coefficients are different for each factor:

$$* \text{Coef. of each Factor} = \frac{\sum \text{Factor analyzed}}{\sum \text{All Factors analyzed}}. \tag{2}$$

The Pugh matrix was implemented to obtain the coefficients, which quantitatively relate each of the studied parameters. The weight of the criteria is given by the technical point of view modelled in GIS, plus the information obtained throughout the research, for example, the conditions mentioned in Table 1 [33].

Phase III, which deals with the application of the geophysical method by SEVs in the selected sites; is done by supplying current to the ground through "electrodes", aligning them symmetrically concerning the center of electrical charge, placing four parallel electrodes, in such a way that the diameter can be increased to at least 100 m; they are placed throughout the study area and are called "current electrodes" (AB), and those that measure the potential difference are called "internal electrodes" (MN).

The application of the method in the field is based on the Schlumberger array to acquire the resistivity of the soil (eqn (3)), which orders the various electrodes in a specific way known as an electrode device. The theoretical foundation is based on the stationary electric field and on detecting the surface effects produced by the passage of electric current to obtain information on its properties.

$$\rho = 2 * \pi * R * (n + 1) * na \tag{3}$$

where ρ: average resistivity at depth (A); R: tellurometer reading in Ohms; N: electrode number; na: distance between electrodes in meters.

With the help of the ABEM Terrameter SAS 1000 equipment, a current is injected in short periods ranging from 0.5 to 0.4 second duration from the external electrodes, and information is collected through the internal electrodes, with a Schlumberger configuration with a range horizontal (AB/2) of 31.6 m, for which the equipment was placed with the established configuration and measurements.

It is necessary to carry out a previous field visit in the selected places to carry out the SEVs. The access of the equipment, excess vegetation, or the material found in the area, be it soil, or rock, can be determined, and avoid concrete. Once compliance with the conditions for the application of the electrical method is verified, a code can be assigned to each of these sites, and the fieldwork is prepared.

The interpretation is performed using the IPI2win software, which approximates the resistivity curve with an error of less than 5%, which increases the confidence level in the results obtained. It must be considered that these are approximations of the lithologies and their depths in the subsoil, a product of the subjectivity of the data since there are no fixed ranges of resistivities. These data tend to vary depending on fractures or water saturation parameters.

The feasibility phase, which usually includes the economic viability analysis, was not carried out because the three selected sites have the same characteristics at the ground and subsoil levels. Therefore, there will be the same expense and preparation in all three before placing the structure.

4 RESULTS

In Phase I, it was possible to identify and analyze the available information with which Figs 3–5 and Table 1 were prepared. In Fig. 3, the four implantation zones found are observed, which delimit the morphological map, such as alluvial colluvium, low hills, medium hills, and populated or non-applicable areas for the implementation of the CSVR. Fig. 4 is the map of the seven delimited areas for the measurement of infiltration. Fig. 5 shows the landslide zone map.

Table 1: Valorization of site selection.

Selected site	1. Aquaculture		2. "Ajá" Park		3. Courts		4. Current collection center		5. Church		6. UPC	
Factor	Sum. influence | Coefficient of each factor											
Population	5	0.14	8	0.21	8	0.22	7	0.19	8	0.21	6	0.17
Hazard (landslides)	8	0.23	8	0.21	9	0.24	7	0.19	7	0.18	7	0.19
Runoff	11	0.31	8	0.21	8	0.22	7	0.19	8	0.21	10	0.28
Topography	6	0.17	8	0.21	7	0.19	5	0.14	7	0.18	8	0.22
Route	5	0.14	7	0.18	5	0.14	11	0.30	8	0.21	5	0.14
Summary	35	1.00	39	1.00	37	1.00	37	1.00	38	1.00	36	1.00
MCP	7.743		7.82		7.649		7.919		7.632		7.611	

Table 2: Results of the SEVs implemented.

Location, and general description	Coordinates and elevation	Resistivity (ω m)	Description by layer
SEV01AC, highly fractured shale outcropping with clay, strike 12°N, located behind UPC (Fig. 6)	616490, 9762175, 85 m	71.6	Sand with gravel thickness of 2.63 m, with a depth of 2.63 m
		380	Consolidated shale (impervious material) with a thickness of 0.854 m, at a depth of 3.48 m
		20.9	Sands with clay, with a thickness of 2.52 m that extends to a depth of 6.01 m
		202	Fractured shale with clay at a depth of 19.8 m, with a thickness of 13.8 m
		1,724	Consolidated rock (fresh shale)
SEV02AC, heading 56°N, located near the ESPOL water intake (Fig. 7)	615823, 9762526, 99 m	79.6	Sand with gravel thickness of 2.32 m, with a depth of 2.32 m
		47.8	Shale fragments with clay with a thickness of 2.63 m, at a depth of 4.95 m
		206	Consolidated shale (impervious material) with a thickness of 15.2 m that extends to a depth of 20.2 m
		9.65	Fine sands with zeolites
SEV03AC, heading 30°N, located by the soccer fields (Fig. 8)	614760, 9763218, 89 m	560	Sand with gravel 0.495 m thick, with a depth of 0.495 m
		12.3	Fine sands with zeolite with a thickness of 0.977 m, at a depth of 1.47 m
		117	Consolidated shales (impervious material) with a thickness of 3.22 m to a depth of 4.7 m
		54.1	Medium grain sands at a depth of 9.49 m, with a thickness of 4.79 m
		109	Consolidated shale with a thickness of 11.4 m, at a depth of 20.9 m
		1,694	Consolidated rock

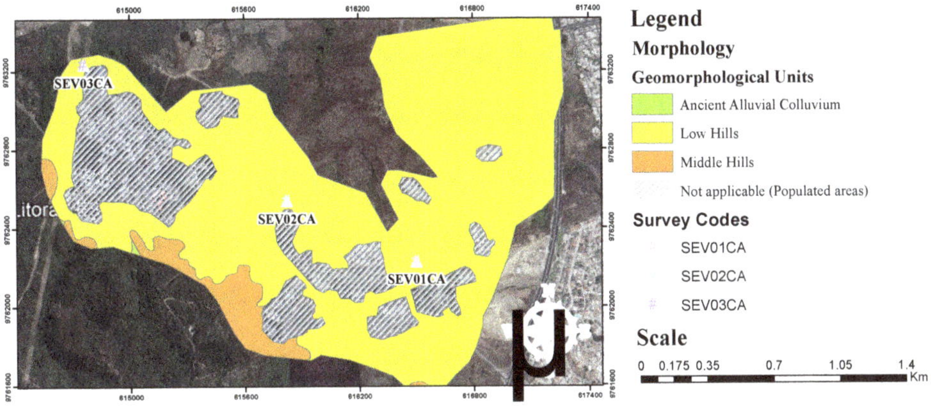

Figure 3: Morphology of the study area.

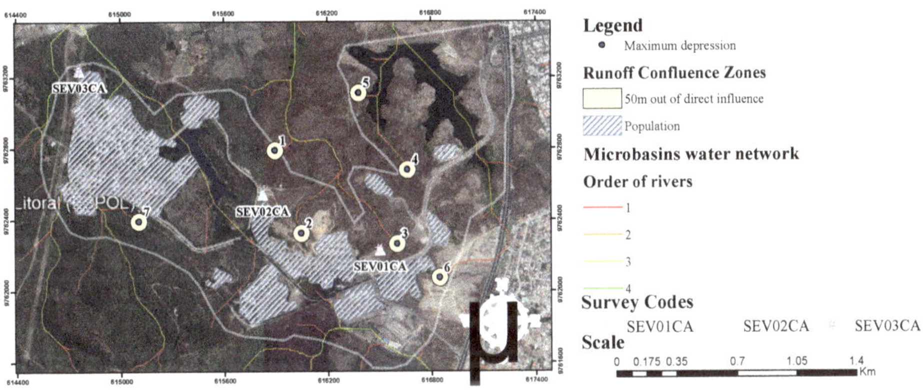

Figure 4: Areas of possible water infiltration.

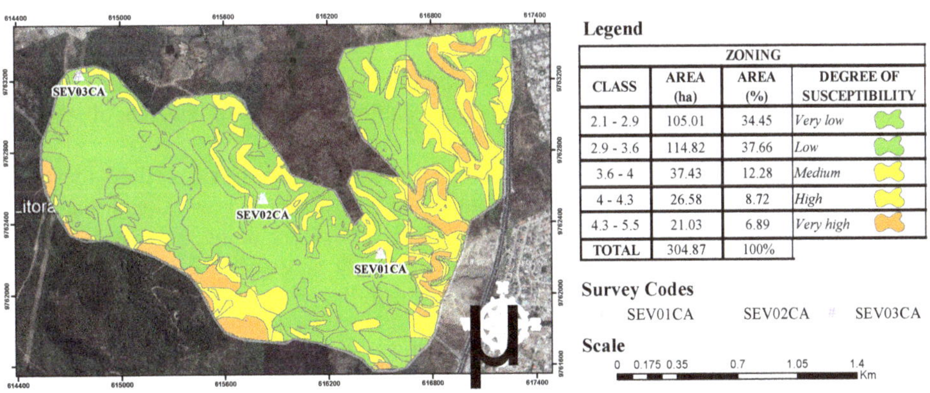

Figure 5: Susceptibility by Mora–Vahrson methodology.

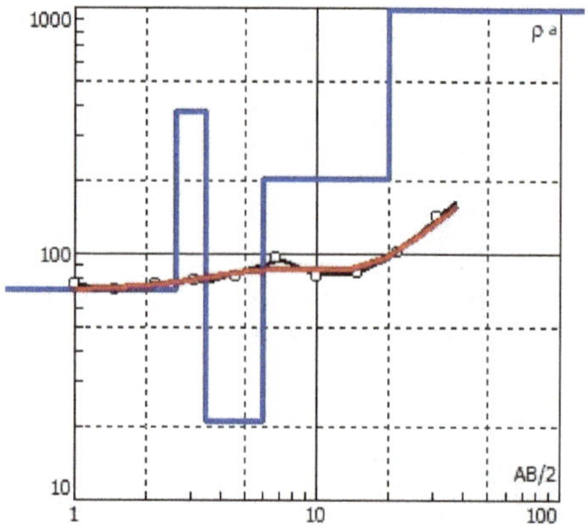

Figure 6: Data processing of point SEV01AC using the IPI2WIN program.

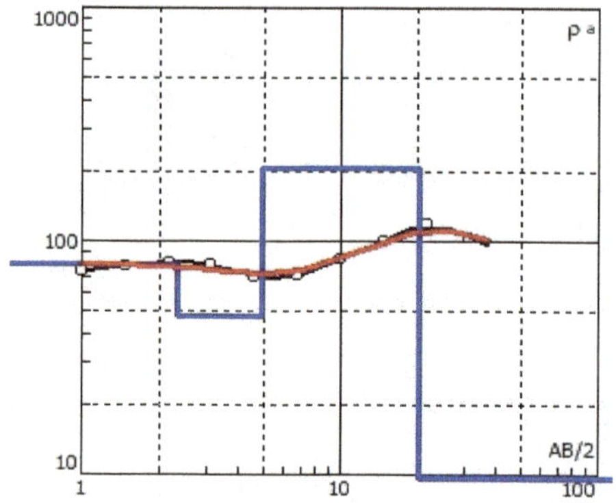

Figure 7: Data processing of point SEV02AC using the IPI2WIN program.

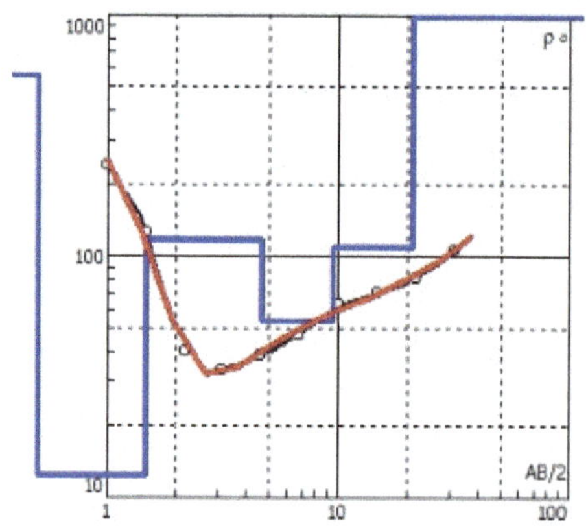

Figure 8: Data processing of point SEV02AC using the IPI2WIN program.

In Phase II, the evaluation of the five factors was carried out with a scale from 1 to 4. The results for each of the six points show in summary Table 1. The three points that obtained the lowest value have selected sites 3, 5 and 6.

In Phase III, which deals with the application of the SEVs geophysical method, these were executed in the three selected points of Phase II (Fig. 3), which now take the names of SEV01AC for point 6, SEV02AC for point 5, and SEV03AC for point 3. Table 2 shows the results. The implemented method achieved a total range of SEVs of 63.2 m, delving into the subsurface approximately one-third of the total range in the horizontal.

5 INTERPRETATION OF RESULTS AND DISCUSSION

The selection of sites where the SEVs were implemented fell on areas whose geological characteristics represent the best conditions to obtain good field data. Under these conditions, the selected places were three (SEV01CA, SEV02CA, and SEV03CA) located on the map (Fig. 1). As seen in Table 1, most of the MCP factors are in the medium to high range (7.611–

7.92). The highest indices are related to the areas where it is difficult to carry out the SEVs and build the CVRS infrastructure, while the lowest values are those that are optimal for carrying out the SEVs and placing the infrastructure. As shown in Kazuva et al. [34], the three places with the best percentage perspective at the time of mapping in GIS were chosen.

The data obtained from GIS and MCP were of great relevance for the reduction of the margin of error when monitoring the possible study sites; these were compared with the article "GIS- and MCD-based suitability assessment for the optimized location of solid waste landfills in Dar es Salaam, Tanzania". The results have a similar form, where various risk factors were also implemented, prioritizing the analytical study for the best approach to the matrix. When analyzing the article "Multicriteria Decision Analysis in Geographic Information Systems for Identifying Ideal Locations for New Substations" and comparing it with our result, it can be explained that the study proposed in both investigations can identify ideal sites, and they are used by the planner to find the best long-term design alternative for work [33].

According to the article "Multi-Criteria Decision Analysis and GIS Approach for Prioritization of Drinking Water Utilities Protection Based on their Vulnerability to Contamination", the values considered are prioritized concerning the weighted criteria based on the objectives expected for the level supply selection. As in this article, the correct sampling of information supported the decisions for the selection of an ideal site [35].

According to the article by Choudhury et al. [36] SEVs are considered for the determination of zones with aquifers directly related to soil and rock, to study the response of the layer and delineate potential zones, as was done in this investigation.

In Ecuador, geosynthetics are widely used in geotechnical engineering, such as slope improvement works, retaining walls, soil improvement systems, sanitary landfills, drainage control and filtration around geotechnical structures [37], [38]. Therefore, in this case it is not about a sanitary landfill but about the construction of the separation, assessment, and recycling center (CVSR).

6 CONCLUSIONS

This research proposed the ideal location for a separation, assessment, and recycling center (CSVR), through the implementation of a decision matrix, with the application of the vertical electrical sounding technique (SEVs). With this technique, the hydrogeological zoning and suitable sites that presented the best conditions were delimited.

From the analysis of the SEVs, it was found that the area is rich in sands, shales, and gravel. Correlating with the geological information of the area, through the interpretation of the field data of the areas where the drillings were carried out, it can be inferred that the contamination of the water resource could not occur. When carrying out the analysis of the three wells, the point SEV01AC is found to be a layer of approximately 3m that extends to a depth of 6.01 m. In the case of clayey material with sand, it is necessary to be careful because, although it is not considered a typical material for an aquifer, due to the presence of sand, it would be a permeable material. Therefore, it would have the possibility of absorbing leachate. Beyond 20 m, a very compact material was found, which decreases the probability of contamination of the water resource.

With all these considerations, it is concluded that the ideal site for the location of the CSVR is the one behind the UPC, at the coordinates (9762175 N, 616490 E) and designated as SEV01AC, which will allow the sustainable management of waste and use of waste at ESPOL.

ACKNOWLEDGEMENTS

We would like to thank Edgar Berrezueta, (Senior Scientist of the Instituto Geológico and Minero de España (IGME)) for his collaboration in the review process of the structure and content of the manuscript, and Ing. Carola Gordillo, Physical Infrastructure Manager of ESPOL – GIF for the support to the executed work.

REFERENCES

[1] Han, H., Zhang, Z. & Xia, S., The crowding-out effects of garbage fees and voluntary source separation programs on waste reduction: Evidence from China. *Sustainability*, **8**(7), 2016. DOI: 10.3390/su8070678.

[2] Colomina, I.A.F., La gestion integral de los residuos sólidos urbanos en el desarrollo sostenible local. **3**, 2005.

[3] Nanda, S. & Berruti, F., Municipal solid waste management and landfilling technologies: a review. *Environ. Chem. Lett.*, **19**(2), 2021. DOI: 10.1007/s10311-020-01100-y.

[4] Kamaruddin, M.A., Yusoff, M.S., Rui, L.M., Isa, A.M., Zawawi, M.H. & Alrozi, R., An overview of municipal solid waste management and landfill leachate treatment: Malaysia and Asian perspectives. *Environ. Sci. Pollut. Res.*, **24**(35), 2017. DOI: 10.1007/s11356-017-0303-9.

[5] Gohain, S.B. & Bordoloi, S., Impact of municipal solid waste disposal on the surface water and sediment of adjoining wetland Deepor Beel in Guwahati, Assam, India. *Environ. Monit. Assess.*, **193**(5), 2021. DOI: 10.1007/s10661-021-09040-y.

[6] Kostakis. I. & Tsagarakis, K.P., Social and economic determinants of materials recycling and circularity in Europe: An empirical investigation. *Ann. Reg. Sci.*, 2021. DOI: 10.1007/s00168-021-01074-x.

[7] Erhardt, T., Garbage in and garbage out? On waste havens in Switzerland. *Environ. Resour. Econ.*, **73**(1), 2019. DOI: 10.1007/s10640-018-0260-x.

[8] Pardo Martínez, C.I. & Piña, W.A., Solid waste management in Bogotá: The role of recycling associations as investigated through SWOT analysis. *Environ. Dev. Sustain.*, **19**(3), 2017. DOI: 10.1007/s10668-016-9782-y.

[9] Merchan, B. et al., Valuation of a community company and its impact on development strategies. pp. 95–107, 2021. DOI: 10.2495/WRM210091.

[10] de Janon, F.V.B. & Neira, P.F.M., Producción de lixiviados, comparación del método de tchobanoglous con experimentos en lisímetros. *Rev. Inst. Investig. Fac. Minas Metal. Cienc. Geográficas*, **21**(42), 2018. DOI: 10.15381/iigeo.v21i42.15789.

[11] Ghaffariraad, M., Ghanbarzadeh Lak, M. & Ebrahimi Sarindizaj, E., Sensitivity analysis of significant parameters affecting landfill leachate generation rate. *Environ. Monit. Assess.*, 194(1), 2021. DOI: 10.1007/s10661-021-09653-3.

[12] Pasalari, H., Farzadkia, M., Gholami, M. & Emamjomeh, M.M., Management of landfill leachate in Iran: Valorization, characteristics, and environmental approaches. *Environ. Chem. Lett.*, **17**(1), 2019. DOI: 10.1007/s10311-018-0804-x.

[13] Demin, A.P., Water resources and food program. *Water Resour.*, **41**(3), 2014. DOI: 10.1134/S0097807814030051.

[14] Bejerano, J.P., Gutiérrez, J.A.G. & Pérez, Y.V., Estimación del caudal medio de lixiviados generados en el vertedero de Viñales, Pinar del Río. **22**, p. 14, 2020.

[15] Carrión-Mero, P., Briones-Bitar, J., Morante-Carballo, F., Stay-Coello, D., Blanco-Torrens, R. & Berrezueta, E., Evaluation of slope stability in an urban area as a basis for territorial planning: A case study. *Appl. Sci.*, **11**(11), p. 5013, 2021. DOI: 10.3390/app11115013.

[16] Manu, E., Agyekum, W.A., Duah, A.A., Tagoe, R. & Preko, K., Application of vertical electrical sounding for groundwater exploration of Cape Coast municipality in the Central Region of Ghana. *Arab. J. Geosci.*, **12**(6), 2019. DOI: 10.1007/s12517-019-4374-4.

[17] Al-Fares, W., Using vertical electrical soundings for characterizing hydrogeological and tectonic settings in Deir El-Adas area, Yarmouk Basin, Syria. *Acta Geophys.*, **64**(3), pp. 610–632, 2016. DOI: 10.1515/acgeo-2016-0025.

[18] Desherevskii, A.V., Modin, I.N. & Sidorin, A.Y., Seasonal variations in specific resistivity in the upper layers of the earth crust. *Seism. Instrum.*, **55**(3), 2019. DOI: 10.3103/S0747923919030058.

[19] Soomro, A., Qureshi, A.L., Jamali, M.A. & Ashraf, A., Groundwater investigation through vertical electrical sounding at hilly area from Nooriabad toward Karachi. *Acta Geophys.*, 67(1), 2019. DOI: 10.1007/s11600-019-00247-9.

[20] Rodríguez, L., Acevedo, R. & Saxton, G., Estudio de puesta a tierra de instalaciones hospitalarias. *Univ. Cienc. Tecnol.*, 13(52), 2009.

[21] Marx-Stoelting, P., Pfeil, R. & Ritz, V., Science-based decision matrix for the identification of endocrine disruptors for regulatory purposes. *J. Für Verbraucherschutz Leb.*, 11(2), 2016. DOI: 10.1007/s00003-016-1016-6.

[22] Carrión-Mero, P., Quiñonez-Barzola, X., Morante-Carballo, F., Montalván, F.J., Herrera-Franco, G. & Plaza-Úbeda, J., Geometric model of a coastal aquifer to promote the sustainable use of water: Manglaralto, Ecuador. p. 17, 2021.

[23] Li, Y., Determining topographic shielding from digital elevation models for cosmogenic nuclide analysis: A GIS model for discrete sample sites. *J. Mt. Sci.*, 15(5), 2018. DOI: 10.1007/s11629-018-4895-4.

[24] Domínguez-Cuesta, M.J., Jiménez-Sánchez, M. & Berrezueta, E., Landslides in the central coalfield (Cantabrian Mountains, NW Spain): Geomorphological features, conditioning factors and methodological implications in susceptibility assessment. *Geomorphology*, **89**(3–4), pp. 358–369, 2007. DOI: 10.1016/j.geomorph.2007.01.004.

[25] Ahmed II, J.B. & Mansor, S., Overview of the application of geospatial technology to groundwater potential mapping in Nigeria. *Arab. J. Geosci.*, **11**(17), 2018. DOI: 10.1007/s12517-018-3852-4.

[26] Merchán-Sanmartin, B. et al., Proposal for the management of solid waste generated in a university campus: A case study, Bilbao, Spain. pp. 593–604, 2021. DOI: 10.2495/SC210481.

[27] Marx-Stoelting, P., Pfeil, R. & Ritz, V., Science-based decision matrix for the identification of endocrine disruptors for regulatory purposes. *J. Für Verbraucherschutz Leb.*, 11(2), 2016. DOI: 10.1007/s00003-016-1016-6.

[28] Sáez, A., Manejo de residuos sólidos en América Latina y el Caribe. p. 16, 2014.

[29] Arias B.L.A., Rivas, E.T. & Vega S.C.N., Propuesta de diseño para un centro integral de reciclaje con inclusión de fuentes alternativas de energía. *Rev. Esc. Adm. Negocios*, **77**, pp. 136–151, 2014. DOI: 10.21158/01208160.n77.2014.820.

[30] Luzieux, L.D.A., Heller, F., Spikings, R., Vallejo, C.F. & Winkler, W., Origin and Cretaceous tectonic history of the coastal Ecuadorian forearc between 1°N and 3°S: Paleomagnetic, radiometric and fossil evidence. *Earth Planet. Sci. Lett.*, **249**(3–4), pp. 400–414, 2006. DOI: 10.1016/j.epsl.2006.07.008.

[31] Berrezueta, E., López, K., González-Menéndez, L., Ordóñez-Casado, B. & Benítez, S., Ophiolitic rocks and plagiorhyolites from SW Ecuador (Cerro San José): Petrology, geochemistry and tectonic setting. *J. Iber. Geol.*, **47**(1), pp. 367–386, 2021. DOI: 10.1007/s41513-020-00154-9.

[32] Sánchez Padilla, C., Lucas, N. & Rivadeneira, J., Levantamiento Geológico ESPOL Campus Gustavo Galindo. 2017. DOI: 10.13140/RG.2.2.12806.27208.

[33] Zambrano-Asanza, S., Chumbi, W.E. & Franco, J.F. & Padilha-Feltrin, A., Multicriteria decision analysis in geographic information systems for identifying ideal locations for new substations. *J. Control Autom. Electr. Syst.*, **32**(5), 2021. DOI: 10.1007/s40313-021-00738-5.

[34] Kazuva, E., Zhang, J., Tong, Z., Liu, X.-P., Memon, S. & Mhache, E., GIS- and MCD-based suitability assessment for optimized location of solid waste landfills in Dar es Salaam, Tanzania. *Environ. Sci. Pollut. Res.*, **28**(9), 2021. DOI: 10.1007/s11356-020-11213-0.

[35] Alvarado, A., Esteller, M.V., Quentin, E. & Expósito, J.L., Multi-criteria decision analysis and GIS approach for prioritization of drinking water utilities protection based on their vulnerability to contamination. *Water Resour. Manag.*, **30**(4), 2016. DOI: 10.1007/s11269-016-1239-4.

[36] Choudhury, J., Kumar, K.L., Nagaiah, E., Sonkamble, S., Ahmed, S. & Kumar, V., Vertical electrical sounding to delineate the potential aquifer zones for drinking water in Niamey city, Niger, Africa. *J. Earth Syst. Sci.*, **126**(6), 2017. DOI: 10.1007/s12040-017-0860-9.

[37] Karpurapu, R., The geosynthetics for sustainable construction of infrastructure projects. *Indian Geotech. J.*, **47**(1), pp. 2–34, 2017. DOI: 10.1007/s40098-016-0215-5.

[38] Bouazza, A. & Zhang, L., Temperature variations of a geomembrane liner in a municipal solid waste landfill from construction to closure. *J. Indian Inst. Sci.*, **101**(4), pp. 725–743, 2021. DOI: 10.1007/s41745-021-00249-x.

MACHINE LEARNING TECHNIQUES APPLIED TO THE VEHICLE ROUTING PROBLEM FOR WASTE SITE INSPECTIONS

FRANCESCO NEBULA, ROBERTO PALUMBO & GABRIELLA GIGANTE
Complex, Intelligent and Autonomous Systems, Italian Aerospace Research Centre (CIRA), Italy

ABSTRACT

During past decades, illegal acts of waste dumping occurred many times in the Campania region of southern Italy. Tons of waste have been dumped in agricultural areas and often illegally burned posing serious health threats to the neighbouring population. Authorities are trying to mitigate such phenomenon by means of continuous monitoring and through actions of remediation. Nevertheless, the number of sites to be visited is often very high with respect to the resources that can be deployed on the territory (such as personnel, patrols, etc.). At the same time, these resources are limited and therefore they should be used in an optimized way. In addition, patrols have to comply with assigned time constraints, considering both the travel time and the time needed to inspect the site. Currently, the tours are assigned by human operators according to their own experience but this solution is highly inefficient. Therefore, a computational tool capable of finding optimal inspection tours in reasonable time is highly desired. For this reason, an appropriate process of risk assessment and priority assignment was set up in order to put in place an optimal resource management from an operational point of view. This work confronts the problem of finding the vehicle routes in a way that a "priority index" is globally maximized. This kind of problem belongs to the class of NP-hard problems and it cannot be solved exactly in practicable time even for a few sites. This paper explores the application of some machine learning techniques to solve the described problem, also performing a comparison with the results obtained using a classical heuristic algorithm taken from literature. In more detail, the use of data clustering and Monte Carlo Tree Search approach is investigated with satisfactory results.

Keywords: vehicle routing problem with profits, clustering, Monte Carlo Tree Search.

1 INTRODUCTION

In the recent years, many illegal waste sites were discovered in the Campania region, in southern Italy. In fact, many agricultural and sub-urban areas are scattered with countless micro-waste dumps. Such dumps typically contain urban, agricultural or even industrial wastes and often flammable materials. These materials may sometimes undergo open-burning because of spontaneous combustion or even because of malicious intents. Open-burnings are a source of cancerogenic and toxic substances, and may cause several diseases in the neighbouring population [1]. Of course, local authorities are trying to mitigate such phenomenon by means of a continuous monitoring of the region and through several actions of remediation.

Within the Crowd for the Environment (C4E) project, several emerging information technologies are integrated in an innovative framework to help local environmental authorities discover and monitor illegal dumping sites. In particular, the detection of the dumping sites comes from spontaneous reports of citizens (that use a specific mobile app for crowd sourcing) and periodic satellite image acquisitions (artificial intelligence extracts areas potentially affected by illegal dumping). Each discovered site is characterized by its spatial position, an estimation of its volume and a rough estimation of the kind waste therein. These kinds of information allow the authorities set up an appropriate process of risk assessment. This process allows to define a "priority" for the inspection of each waste site (which is

WIT Transactions on Ecology and the Environment, Vol 257,
www.witpress.com, ISSN 1743-3541 (on-line)
doi:10.2495/WMEI220121

related to the potential danger posed by the site) and then arrange an inspection campaign to assess the validity of the site report and to decide the remediation actions needed.

Actually, one of the objectives of the C4E project is to help local authorities organize the on-field inspections of illegal dump sites.

Currently, the tour assignment to each patrol is provided by human operators only, based on their own experience. Unfortunately, this method is highly inefficient being slow and inaccurate. For these reasons, it is strongly desired to determine the tour assignments automatically with the help of a computational tool. Consequently, our problem consists in finding a computational procedure that determines the vehicles routes in such a way that an appropriate index – hereafter referred as the "overall collected priority" – is maximized, also considering some time constraints like, for example, the work shifts of the operators. This problem belongs to the class of the so-called vehicle routing problems (VRPs) and is generally very hard to solve, even for few numbers of nodes (NP-hard problems).

In a previous work [2] a heuristic algorithm from literature [3] was successfully applied to solve the described problem in a similar operative scenario. However, in this work, we explore an alternative approach leveraging on machine learning (ML) techniques, comparing its outcomes to the same heuristics.

This paper is organized as follows. Section 2 reports the description of the problem to be solved. Section 3 describes the methodology and tools used to build the proposed solution. Section 4 describes the procedure followed to evaluate the performances of the proposed solution, and the main results are shown. Finally, Section 5 collects the main conclusions of this work.

2 PROBLEM STATEMENT

2.1 Background

VRP is one of the most studied combinatorial optimization tasks [4]. The attention paid for VRP, along with its many variants, is justified by its many applications in transportation and logistics, vital fields of the modern economy. Because of its practical relevance, there is a strong interest in finding new solution approaches, despite the existence of several heuristic and approximate methods.

Generally speaking, in a VRP, a fleet of vehicles can visit a set of customers, achieving the fulfilment of a specific service. The objective is to schedule the visits in a way that the use of the available resources is optimized, with respect to some given constraints.

In particular, in the VRP with Profits (VRPP) – also known as Team Orienteering Problem (TOP) [5] – the overall collected profits must be maximized, under some predefined constraints (time deadlines, vehicles capacities, etc.) that make usually impossible to visit all customers using one patrol only. Consequently, both the order of visits and the subset of nodes to be visited must be found.

2.2 Approach

For our specific scenario, the following two-steps procedure is implemented. Firstly, the illegal waste site reports for each municipality of the Campania region are initially grouped in some geographical zones (clusters) in order to satisfy a given criterion based, for example, on the maximum admissible visit time. Each cluster is then characterized by three features:

- a "centroid", defined as the cluster "centre of gravity" and representing the above-mentioned "customer";

- a "cumulative" priority, defined as the average of the priorities of the grouped micro-dumps, equal to their volumes weighted by the dangerousness of the contained materials;
- a "cumulative" collection time, sum of two different terms: the minimum travel time to visit (only once) all the sites with a fixed speed V (travelling salesman problem (TSP)) and the sum of times needed to inspect them. Hereafter this total time will be referred as "the time to collect" the cluster priority.

In the second phase, the ML solver is applied on the obtained set of clusters for each municipality. This solution is then compared to a baseline solution, described in Butt and Cavalier [3] and known under the acronym of MAXIMP, for validation purposes. Finally, when the above described strategic step is concluded, all the clusters within each municipality can be "tactically" visited, using the TSP paths computed in the previous step. Fig. 1 shows the described approach.

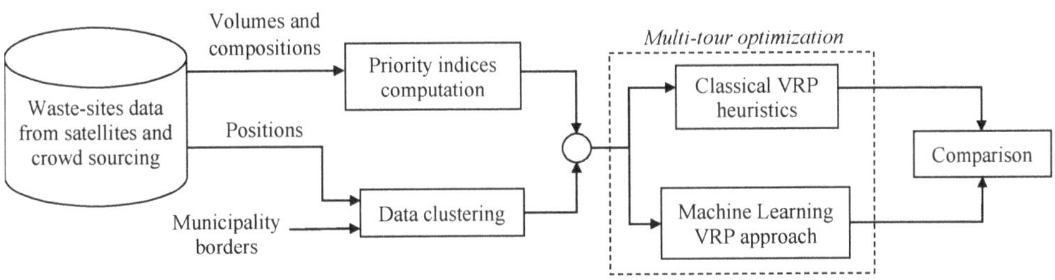

Figure 1: The overall approach followed in this work.

2.3 Formulation

Hereafter we formalize the strategic problem described in Section 2.2 for homogeneous vehicles starting from a unique depot (0) because the general case can be always decomposed in similar problems.

Let us define a complete graph $G=[N,A]$ where $N=\{0, 1, \ldots, n\}$ is the set of centroids and $A=\{a_{ij}\}$ is the set of the arcs between the i-th and j-th centroids ($i,j \in N-\{0\}$). We also define the following quantities:

- the distance matrix $\{d_{ij}\}$, composed by the road distances between the i-th and the j-th centroids, generally asymmetric, i.e., $\{d_{ij}\}\neq\{d_{ji}\}$;
- the priority p_i ($p_0=0$), i.e., the cluster volume, weighted by the harmfulness of the contained waste;
- the time b_i required to collect p_i ($b_0=0$), i.e., the time required for the cluster inspection.

Moreover, a tour is defined as "feasible" through a subset of G if it starts and ends in the depot, if it visits each centroid not more than once and if it is completed within a given maximum time T. Using this definition, our problem is to find not more than m feasible tours such that the overall priorities, collected over all tours, is maximized.

3 TOOLS

3.1 Geographic information systems

When dealing with spatial data and their relations, it is often necessary to analyse, manipulate and process spatial information. This can be carried out using geographic information systems (GIS). Specifically, we used the open-source application QGIS [6] together with Python scripts to:

- visualize the micro-dump sites on the map;
- analyse the characteristics of the sites, in terms of volume and composition;
- visualize the borders of the municipalities;
- compute the number of sites belonging to each municipality;
- compute the road distances between two selected points on the map.

To compute the road distances between the micro-dumps, it is required to know the road graph. The related GIS-readable shapefile was downloaded from OpenStreetMap [7] using a Python library named OSMnx [8]. Next, using the QGIS plug-in called QNEAT3 (QGIS Network Analysis Toolbox 3 [9]), it was possible to compute the matrix of the road distances. Fig. 2 reports the road graph for the area of interest (long./lat. intervals: 13.9°E–14.5°E, 40.79°N–41.15°N).

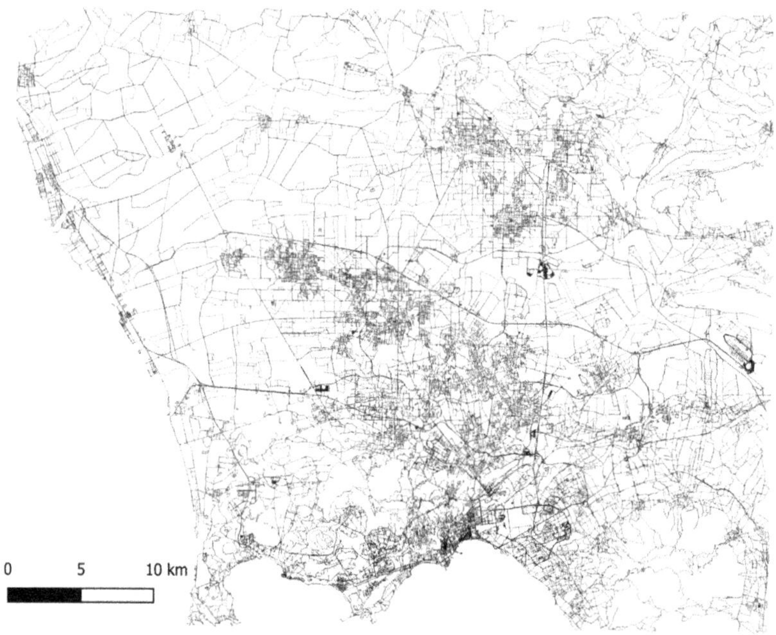

Figure 2: Road graph from OpenStreetMap [7].

It is worth to note that the road distance matrices are generally asymmetric because the connection between two points cannot be always achieved using two-ways roads and a longer path must then be followed. Consequently, points that apparently seem to be "close to each other", could be very far apart when considering road distances.

3.2 HDBSCAN

Clustering algorithms are used to explore and analyse patterns of similarity in a dataset. Centre-based algorithms find only convex shapes with circular symmetry: they are a bad choice for finding clusters of arbitrary shapes. Instead, density-based clustering is a paradigm where clusters are identified as data partitions with a higher density than their surroundings and therefore they are more suitable for finding clusters of generic shapes. Objects that do not meet a certain density criterion are discarded as "noise".

In addition, partitioning methods group items into a predefined number of categories, while hierarchical algorithms build a hierarchy of data partitions in order to reveal the "intrinsic" data structure (so there is no need to specify the number of categories).

For the proposed problem, and in general for spatial data mining, it is desirable to obtain clusters with arbitrary shapes (in particular long-shaped clusters, due to the fact that micro-dumps usually lie along roads), with varying densities and also without specifying their number.

To this end, a suitable choice appears to be the HDBSCAN algorithm [10], a "Hierarchical" upgrade of DBSCAN (density-based spatial clustering of applications with noise). It is a density-based clustering algorithm, that builds a hierarchical cluster tree (dendrogram), from which the clusters are extracted. In this process, the algorithm is allowed to decide how many clusters to create, using an internal stability measurement.

More in detail, in HDBSCAN the dendrogram is "condensed" applying different cuts at different heights (densities), minimizing the points that are not falling in any cluster: this results in a more compact tree with fewer clusters that lose points and it will be used to select clusters more stable and persistent [11]. This process allows the tree to be cut at varying height, picking our varying-density clusters based on cluster stability.

HDBSCAN has two main parameters:

- "minimum cluster size", imposing the minimum cluster dimension;
- "min samples", less intuitive than the previous one, it represents "how much noise" one wants to exclude in the cluster generation and, consequently, "how many big spatial structures" one wants to consider in the clustering process.

3.3 Monte Carlo Tree Search with UCB-based policy

The Monte Carlo Tree Search (MCTS) [12] is a heuristic search algorithm for some specific decision processes, especially employed in software that play board games: in this context, MCTS is used to solve the game tree, with the notable advantage that only the game mechanics needs to be implemented. In fact, MCTS can theoretically be applied to any domain that can be described by "state and action" pairs and by a simulation to forecast the game outcome. MCTS is typically employed using a tree policy that is based on the upper confidence bound (UCB) [13] formula applied to trees (known as UCT). This approach is in fact effective in solving the "exploration/exploitation dilemma" allowing the algorithm to converge more easily to suboptimal solutions or better.

In this paper, we leverage on the computational power of MCTS to solve the VRP described in Section 2. To this aim, the most critical issue is to map this kind of problem into a game tree formalism. First of all, the MCTS is treated as single-player. The root node of the tree is the starting point of the route namely the depot. Each tree node is then a sub-sequence of visited centroids starting from the depot. Actions from each tree node are all the unvisited centroids that allow the return to depot within the admissible time window. The MCTS selection step is carried out using the abovementioned UCT policy. The rollout step is carried out using a random policy among the possible valid actions. As explained in Section 2, the VRP solution is represented by a sequence of visited centroids that maximizes the collected priority within a predefined amount of time. Therefore, in the MCTS context, the tree game ends as soon as there is no more time left to visit new nodes other than going back to the depot. When this condition is met, the node is considered terminal and a possible solution is found. In order for the UCT to expand the tree in the right directions, solutions must be properly classified as good/poor or promising/unpromising. The reward function and backpropagation step favour only nodes that yield solutions that maximize the collected priority within the time constraint. Following Mańdziuk and Nejman [14] we have also given importance to solutions that appear promising, rewarding – in a linear fashion – solutions for which the resulting collected priority falls in the interval $[\alpha * best\_priority, best\_priority)$ where α is a parameter less than 1 and set to 0.5, and *best_priority* is the current best collected priority found.

4 RESULTS

4.1 Raw input data

The data considered in this work are inside a specific area of the Campania region (Italy) particularly affected by the problem of the open-burning of waste. This area is identified in Fig. 3 by the dashed line (longitude/latitude intervals: 13.90°E–14.50°E, 40.79°N–41.15°N). The selected area is a rectangle of 51 km × 41 km and, although it represents a small area with respect to the overall Campania region, it contains 66% of the reported micro-dumps (4397 over 6629).

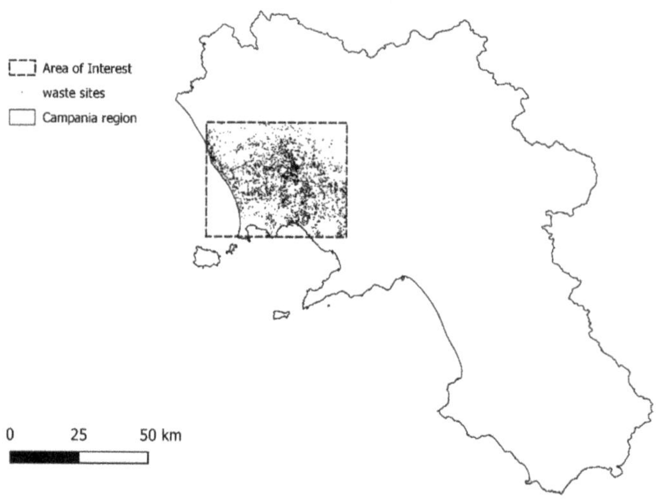

Figure 3: Area selected for the waste sites inspections within the Campania region (Italy).

4.2 Municipalities selection

A geo-tessellation of the Campania region is performed using the borders of the local administrative divisions (Municipalities). The reason of this grouping is that the local patrols are allowed to visit only sites included in their Municipalities.

Furthermore, Municipalities are selected considering the number of the waste-site reports. The areas with less than 20 points are filtered out because in these cases clustering does not confer any significant advantage (Fig. 4).

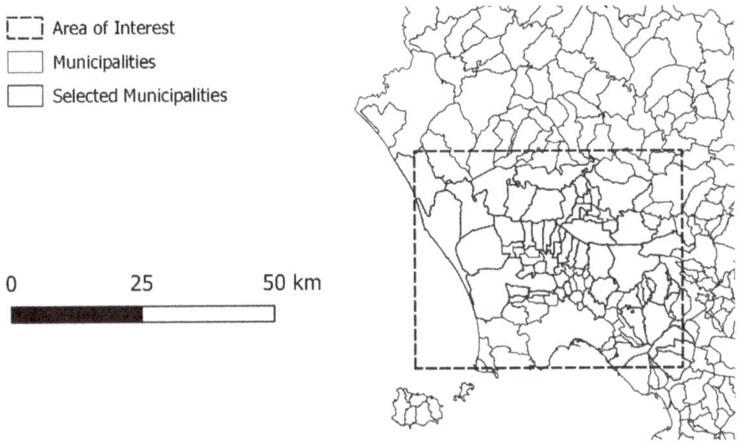

Figure 4: Selected Municipalities.

For example, Fig. 5 shows one of the selected municipalities (Villa Literno), its 204 waste sites (points) and the depot (diamond). The roads are also shown and it can be clearly seen that the waste sites are often located along the roads.

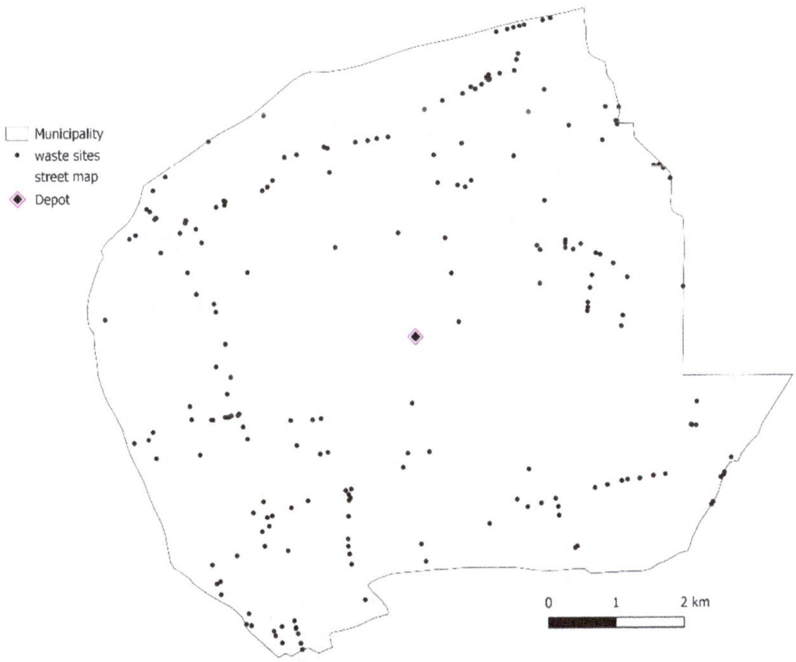

Figure 5: Waste sites, depot and roads (municipality of Villa Literno).

4.3 Clustering analysis

For each selected municipality, the set of waste sites is processed using the HDBSCAN clustering algorithm described in Section 3.2, considering the road distances instead of the line-of-sight (Euclidean) ones.

In this process, an operative requirement is imposed: the total inspection time for each cluster must be the *max*(45 min, 2**k* min) where k is the number of waste sites within the cluster. Imposing this constraint, the two HDBSCAN parameters are chosen consequently.

In Fig. 6 the results of the HDBSCAN clustering are shown for San Tammaro municipality. The points not included in any cluster represent the "noise", i.e., not considered in the outcome of the algorithm. From an operational perspective, these points can be later re-analysed repeating the same two-step procedure described in Section 2.2.

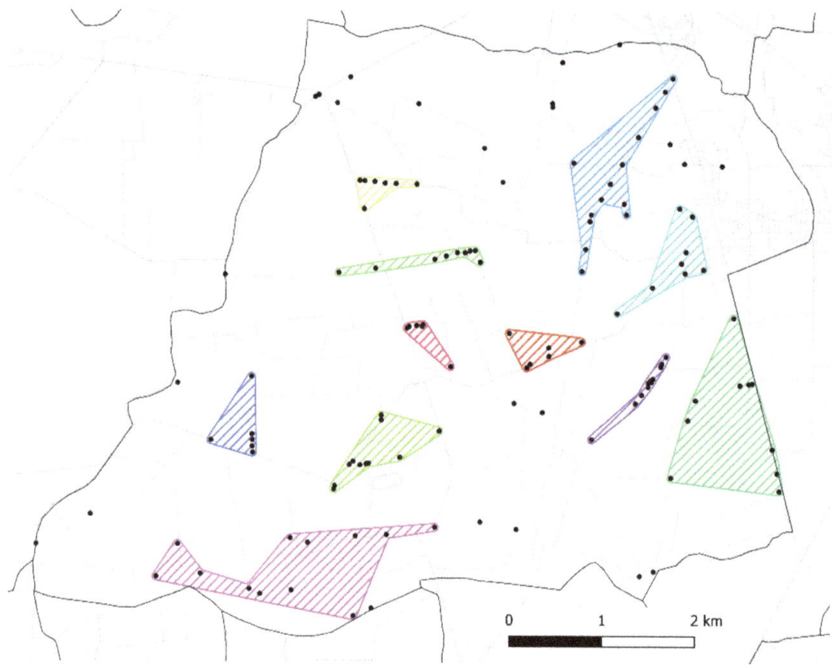

Figure 6: HDBSCAN clustering (municipality of San Tammaro).

As already noted, the obtained clusters are generally long-shaped, reflecting the fact that the micro-dumps are formed mainly along the roads and to group them in this way leads obvious practical advantages for the operators.

In Table 1 some relevant data about the clustering process are reported:

- Municipality ID;
- Name of municipality;
- Total number of the waste sites within the municipality borders;
- Number of clusters;
- Priorities of clusters;
- Percentage of the unclassified waste sites.

Table 1: Clustering data.

ID	Municipality	Total waste sites	Clusters	Cluster priorities (average volume, m^3)	Unclassified waste sites (%)
1	Casal di Principe	95	9	4, 8, 2, 87, 2, 12, 9, 20, 16	27
2	Teverola	37	11	3, 2, 12, 10, 2, 2, 5, 2, 13, 3, 1	11
3	Afragola	131	9	8, 3, 6, 6, 15, 3, 10, 5, 8	25
4	Sant'Antimo	46	9	2, 3, 2, 4, 1, 3, 2, 3, 8	4
5	Casaluce	48	11	4, 9, 4, 2, 9, 3, 140, 8, 8, 4, 4	15
6	Somma Vesuviana	48	6	8, 4, 12, 4, 11, 4	8
7	Castel Volturno	235	11	4, 4, 6, 13, 11, 5, 32, 10, 6, 3, 9	22
8	Macerata Campania	66	10	10, 4, 8, 4, 6, 4, 4, 7, 7, 6	9
9	Parete	21	7	3, 9, 4, 13, 5, 3, 12	19
10	Casoria	40	11	4, 3, 4, 4, 7, 2, 6, 2, 4, 18, 2	15
11	Carinaro	53	7	2, 2, 2, 5, 2, 8, 6	32
12	Succivo	72	11	3, 2, 2, 2, 3, 5, 2, 2, 4, 6, 6	17
13	Casandrino	27	9	10, 2, 3, 5, 2, 6, 4, 4, 2	11
14	Santa Maria Capua Vetere	84	8	2, 10, 6, 10, 5, 8, 5, 4	23
15	Grazzanise	53	9	14, 7, 10, 4, 18, 2, 101, 2, 30	17
16	Villa di Briano	44	8	9, 5, 20, 12, 7, 7, 8, 6	16
17	San Felice a Cancello	29	9	5, 5, 2, 4, 2, 3, 2, 8, 4	17
18	Cancello ed Arnone	35	9	7, 6, 59, 2, 8, 2, 4, 15, 2	17
19	Aversa	24	8	12, 2, 12, 18, 2, 2, 5, 16	4
20	Castello di Cisterna	20	5	8, 10, 34, 15, 8	30
21	Terzigno	36	10	11, 3, 4, 8, 5, 6, 2, 2, 5, 11	11
22	San Vitaliano	23	6	13, 4, 7, 8, 5, 2	17
23	Mondragone	99	11	20, 8, 30, 9, 4, 12, 4, 3, 42, 14, 3	15
24	Ercolano	28	8	3, 8, 2, 8, 3, 2, 4, 4	14
25	San Cipriano d'Aversa	29	9	5, 8, 13, 2, 10, 2, 9, 3, 51	14
26	San Prisco	29	8	16, 4, 4, 5, 11, 52, 4, 6	14
27	Giugliano in Campania	345	11	4, 4, 6, 5, 5, 4, 7, 6, 14, 6, 5	21
28	Caserta	54	10	5, 2, 5, 2, 2, 3, 2, 1, 3, 7	20
29	Qualiano	30	9	252, 3, 2, 7, 1, 2, 2, 10, 1	20
30	Ottaviano	32	5	12, 7, 6, 2, 12	9
31	Santa Maria la Fossa	42	7	1, 2, 21, 26, 8, 4, 3	26
32	Villa Literno	204	11	12, 11, 7, 36, 5, 11, 3, 2, 7, 5, 5	31
33	Orta di Atella	92	10	7, 23, 11, 5, 85, 21, 20, 3, 4, 4	33
34	San Tammaro	125	11	6, 3, 7, 5, 3, 3, 8, 3, 7, 5, 4	20
35	Frignano	62	11	3, 2, 13, 9, 11, 5, 2, 12, 7, 11, 1	10
36	Caivano	139	11	6, 6, 7, 15, 10, 11, 6, 8, 5, 24, 31	19
37	Marcianise	217	10	5, 2, 3, 3, 4, 3, 3, 3, 7, 11	39
38	Marigliano	102	10	8, 5, 7, 6, 7, 14, 7, 6, 10, 5	28
39	Acerra	203	9	8, 16, 14, 10, 15, 9, 7, 4, 3	23
40	Gricignano di Aversa	106	8	3, 4, 3, 6, 2, 10, 4, 9	24
41	Capua	60	11	7, 5, 3, 8, 9, 4, 13, 10, 7, 12, 4	12
42	Sant'Anastasia	61	6	4, 20, 6, 4, 32, 11	13
43	Villaricca	26	7	4, 7, 5, 5, 1, 16, 3	12

4.4 UCT performance evaluation

In this section, UCT performance will be compared to the results of the above-cited MAXIMP algorithm, which is used in the same scenario conditions, selecting an average value of 0.5 for its α parameter (see [3]). UCT performance cannot be compared to the exact optimal solutions as the time required for exhaustive route enumeration is not feasible even for few nodes (for example, just for 9 nodes, 9! = 362880 permutations imply a computational time of about 20 min, with a state-of-art personal computer). Instead, the computational time required by MAXIMP is estimated below one second.

Finally, in our scenario, the UCT computational time, although around an average value of about 70s, is fully compliant with an operational usage on a daily basis.

The problem settings are hereafter reported:

- T = 8 hours, typical Italian work shift;
- m = 3, maximum number of the vehicles starting from the same unique depot;
- V = 20 km/h, constant vehicles speed;
- b_i = 2 min equal for each i-th site, average time needed to perform a site inspection.

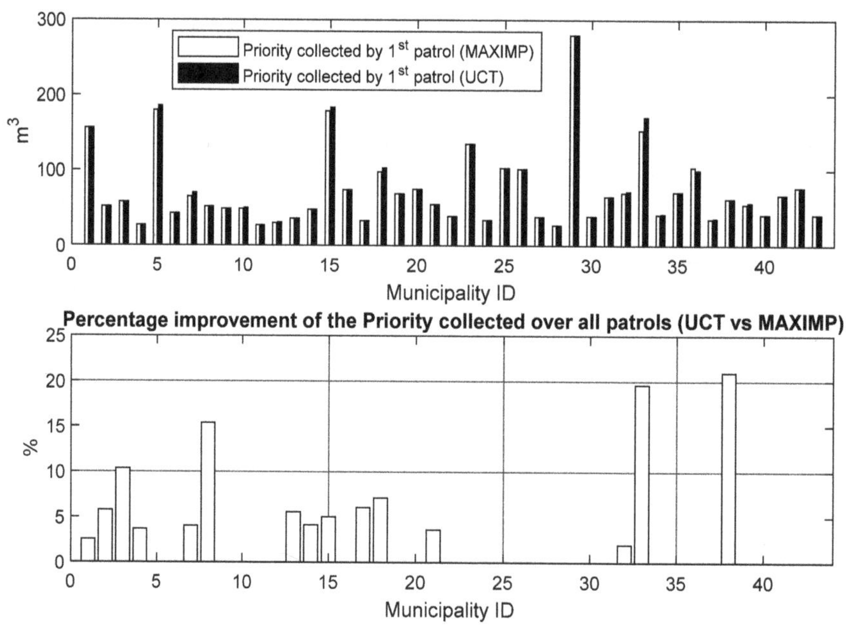

Figure 7: Collected priorities by MAXIMP and UCT (1st patrol only) and UCT improvement.

In Fig. 7 two different information are shown:

- the top chart shows the collected priority (the average cluster volume weighted with the materials dangerousness) of the first patrol only, as planned by both MAXIMP and UTC. It is possible to see that, already using only one patrol, UCT always outperforms the first one, except for the municipality number 36 “Frignano” (104 vs 100).

- the bottom chart shows the percentage improvement of the overall collected priority by UCT with respect to MAXIMP (+2.7% mean, 5.1% std), using all patrols and over all the municipalities. It should be noted that the overall priority collected using UCT always matches the theoretical one (sum of the priority values per municipality (see Table 1)).

For the first patrol only, the mean and the standard deviation values of the mean residual time with respect to the deadline T, are [0.79 0.64] hours and [0.41 0.62] hours, respectively for MAXIMP and UCT: this means that the UCT generally exploits better the available time interval T.

In terms of the number of the patrols involved, MAXIMP often stops when few sites are left, while UCT always uses the same number or more. However, for the second or third patrol, the solution is sometimes trivial, e.g., only one or two sites remain to be visited (Fig. 8).

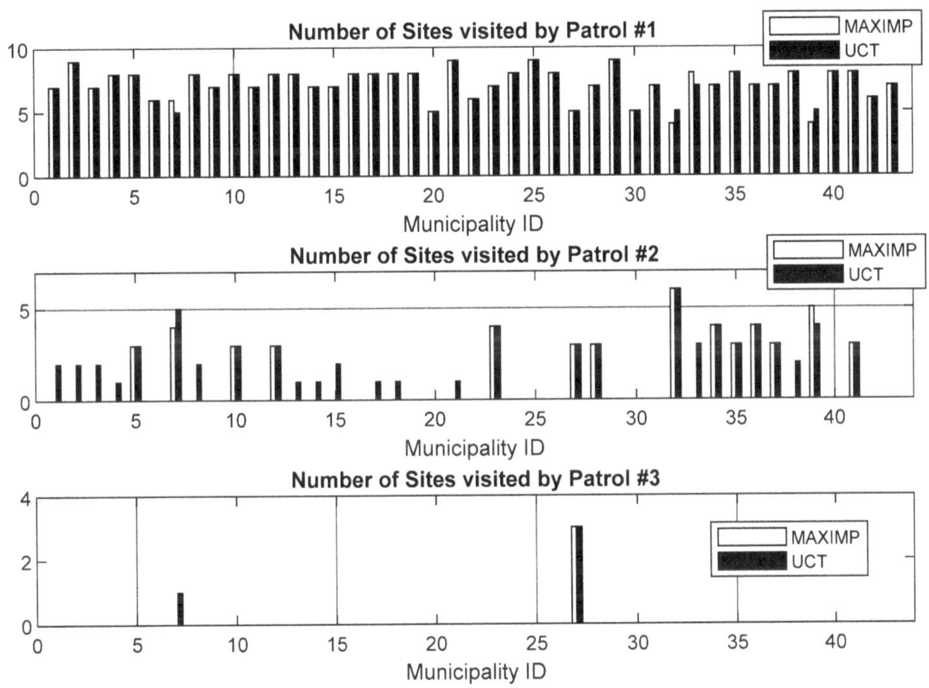

Figure 8: Number of waste sites visited by each patrol (MAXIMP and UCT comparison).

5 CONCLUSIONS

In this work some machine learning (clustering and Monte Carlo Tree Search) techniques are applied to optimize the use of resources for waste-sites inspections. More in detail, these techniques are used to perform an optimal multiple-tour assignment for the available patrols in order to globally maximize the number of visited sites that have the highest priority (which is related to their volume and the dangerousness of their waste).

The solutions obtained using the Monte Carlo Tree Search with upper confidence bounds (UCT) are compared to those obtained using a heuristic algorithm taken from literature (MAXIMP), often used to solve this kind of VRP. For the specific scenario here considered, the overall priority collected using UCT always matches or outperforms MAXIMP, with a better exploitation of the time interval given for the inspections. In addition, the UCT computational time (~1 min) enables also practical applications.

Besides the demonstrated performance, the other advantage of the presented approach is the fact that, once the considered problem is mapped into a game tree formalism, the UCT can operate effectively without any knowledge of the particular domain, apart from the game rules and the end conditions. Possible future improvements may focus on investigating modifications to the tree policy in the MCTS algorithm. This may improve its performance in finding optimal solutions as the state space grows (graphs with more nodes).

ACKNOWLEDGEMENTS

The proposed work has been developed within the Italian national funded project C4E "Crowd for Environment" from 2018 to 2021, aiming to develop a support decision system for the local authorities in the monitoring and removal of illegal waste. Founded by Italian Ministry of Research Grant Number RNA-COR 896199.

REFERENCES

[1] Mazza, A., Piscitelli, P., Neglia, C., Rosa, G. & Iannuzzi, L., Illegal dumping of toxic waste and its effect on human health in Campania, Italy. *International Journal of Environmental Research and Public Health*, **12**(6), pp. 6818–6831, 2015.

[2] Nebula, F., Gargiulo, F., Gigante, G., Pascarella, D. & Cicala, L., Multiple-tour constrained optimization for waste sites inspections. *International Conference on Optimization and Learning – OLA2021,* pp. 93–95, 2021.

[3] Butt, S.E. & Cavalier, T.M., A heuristic for the multiple tour maximum collection problem. *Computers and Operations Research,* **21**, pp. 101–111, 1994.

[4] Toth, P. & Vigo, D., An overview of vehicle routing problems. *The Vehicle Routing Problem*, eds P. Toth & D. Vigo, SIAM: Philadelphia, pp. 1–26, 2002.

[5] Vansteenwegen, P. & Souffriau, W., The orienteering problem: A survey. *European Journal of Operational Research*, **209**, pp. 1–10, 2011.

[6] QGIS Geographic Information System, QGIS Development Team. www.qgis.org.

[7] OpenStreetMap, OpenStreetMap contributors. www.openstreetmap.org. Accessed on: 18 Sep. 2021.

[8] Boeing, G., OSMnx: New methods for acquiring, constructing, analyzing, and visualizing complex street networks. *Computers, Environment and Urban Systems*, **65**, pp. 126–139, 2017.

[9] QNEAT3, QGIS Network Analysis Toolbox 3. root676.github.io. Accessed on: 28 Feb. 2022.

[10] McInnes, L., Healy, J., Astels, S., HDBSCAN: Hierarchical density based clustering. *Journal of Open Source Software*, **2**(11), 2017.

[11] McInnes, L. & Healy, J., Accelerated hierarchical density clustering. *2017 IEEE International Conference on Data Mining Workshops (ICDMW)*, pp. 33–42, 2017.

[12] Browne, C., Powley, E., Whitehouse, D., Lucas, S., Cowling, P.I., Rohlfshagen, P., Tavener, S., Perez, D., Samothrakis, S. & Colton, S., A survey of Monte Carlo Tree Search methods. *IEEE Transactions on Computational Intelligence and AI in Games*, **4**(1), pp. 1–43, 2012.

[13] Auer, P., Cesa-Bianchi, N. & Fischer, P., Finite-time analysis of the multiarmed bandit problem. *Machine Learning*, **47**, pp. 235–256, 2002.

[14] Mańdziuk, J. & Nejman, C., UCT-based approach to capacitated vehicle routing problem. *Artificial Intelligence and Soft Computing*, pp. 679–690, 2015.

BUILDING COMMUNITY TRUST IN THE PROCESS OF ESTABLISHING A LOW- AND INTERMEDIATE-LEVEL RADIOACTIVE WASTE STORAGE FACILITY: THE CASE OF CROATIA

GERAN-MARKO MILETIĆ, MATEA MILAK & MATEO ŽANIĆ
Institute of Social Sciences Ivo Pilar, Croatia

ABSTRACT

In Croatia, the radioactive waste management programme foresees the construction of a national central storage for the management of both institutional radioactive waste as well as low and intermediate-level radioactive waste from the Krško Nuclear Power Plant. According to preliminary procedures, the former military logistic complex in the municipality of Dvor was identified as a single preferred location for the future storage facility. Currently, that location is the subject of detailed field testing and impact analysis. One of the conducted impact studies focused on the perception of the challenges and opportunities for the local community regarding siting the radioactive waste storage in its municipality. A broader analysis of social acceptance of technical risks included the issue of community trust, and in this paper we will focus on that aspect of the problem. More precisely, we were interested in the mechanism of building community trust in the state authorities in charge of the implementation of the national radioactive waste management programme. To better understand this, we relied on the concept of social licence to operate. This concept emphasises a relationship between community trust-building and the perception of justice of the project rationale regarding both its procedural and distributive components. This paper aims to investigate the existence of the indicated relationship in a situation where the decision-making process for the site selection is still ongoing. The data for this analysis were collected in 2020 through a questionnaire from 424 respondents living in the municipality of Dvor. The results of the analysis revealed that elements of both procedural and distributive justice were statistically significant predictors of the level of community trust. However, expectations regarding a fair procedure have a somewhat stronger effect on the trust level than expectations that the project would bring benefits to the community.

Keywords: community trust, radioactive waste management, procedural justice, distributive justice, social licence to operate, Croatia.

1 INTRODUCTION

Radioactive waste (RAW) is one of the by-products of industrial–technological development, and its disposal still poses a great challenge to modern societies [1]. It is a waste generated by human activity in various sectors, such as energy, medicine, science and agriculture, and due to its characteristics it belongs to the category of hazardous substances which can have an adverse effect on the environment and human health. These adverse effects are associated with the amount of ionizing radiation, and in this sense we distinguish low and intermediate level RAW from high level RAW. While efforts to establish a safe model for the disposal of high-level RAW is still underway, there is a consensus among experts that appropriate technical solutions and operational procedures can reduce the risk of low and intermediate level RAW to the level of negligible risk, while a dozen facilities for the management of such waste have already been established in various parts of the world [2]. Despite scientific optimism about the management of low and intermediate level RAW, the general public remains inclined to attribute significant individual, group and economic risks, as well as environmental risks, to such infrastructure [3].

www.witpress.com, ISSN 1743-3541 (on-line)
doi:10.2495/WMEI220131

The mentioned controversies are especially evident during the discussions concerning the issue of siting RAW management infrastructure [4]. This is not surprising, since centralised management of RAW is from a national perspective safer, more environmentally friendly, more efficient and ultimately cheaper. However, from the perspective of the local community in which such infrastructure is located, this solution is perceived as extremely burdensome [5], [6]. In addition to the indicated imbalance in the distribution of environmental burdens and benefits, doubts about the reasonability and validity of the decision on the place and manner of RAW management is another cause for opposition which often occurs in the local community. The antagonism towards the idea of storing of RAW in one's own backyard is further intensified by the decline in trust in public institutions, which has been observed over a longer period even in the developed countries of the world [7]. Due to the complexity and long timescale of the RAW disposal process, the state is the main guarantor not only for using adequate technical solutions but also for respecting administrative and operational procedures and ensuring that the local community would not be deceived in such a big game. In this context, a lack of confidence in the effectiveness of the state apparatus can further complicate the establishment of RAW management infrastructure [8].

Therefore, the issue of building trust remains key to success in ensuring the coexistence of the local community and the infrastructure for RAW management. Having this in mind, our paper focuses on the issue of building the local community's trust in the state authorities responsible for RAW management in Croatia, where the establishment of a national RAW management centre is underway. The paper is organized as follows: Section 2 provides a literature review on the relationship between the perceived justice, the trust in key stakeholders and the attitude of the local community towards the idea of siting an infrastructure for RAW management in their local area. Background information about the preferred location and the situation concerning the RAW storage facility development in Croatia were introduced in the Section 3. Section 4 describes the research method and the survey instruments. The presentation of the analysis results is provided in Section 5. The paper is concluded with a discussion of the key findings.

2 LITERATURE REVIEW

Over the years, there have been some changes in framing the radioactive waste problems, starting from viewing only technical problems to including the social dimension of acceptance and legitimacy. As Bergmans et. al [4] state, more attention has been devoted to legitimacy and participation in the siting procedures. According to Cotton [9], socio–technical reframing occurred, which includes integration of the community's and stakeholders' values in the governmental decision-making process.

Bergmans et al. [4] state three main benefits of citizens' participation in decision-making process when it comes to controversial risk issues: (1) it leads to increased legitimacy of decision outcomes, (2) an enhanced knowledge base produces better decisions, and (3) it leads to a more democratic society by securing democratic values. Today, participation is included in the laws and regulations in many countries, with the aim of producing a dialogue among all actors, which can lead to mutual understanding, trust, and cooperation [10].

According to Wolsink [11], community acceptance of siting decisions has a temporal dimension and follows a U-curve; going from a high level of acceptance to a typically low level of acceptance in the siting phase, and looping back to a higher level in the phase once the project is established. The whole process of acceptance is affected and constructed by a dynamic interaction between three main actors: community, stakeholders and companies. Wüstenhagen et al. [12] pointed out that the perceived justice of procedures and outcomes are crucial in gaining a community's acceptance for a new project. According to Krütli et al.

[13], [14] studies on residents' perception of justice in the case of a nuclear waste repository in Switzerland, the existence of procedural justice affected the decision of acceptance more than distributive justice did.

Vilhunen et al. [15] studied the perception of justice and trust among the residents of two nuclear communities in Finland. The main results showed that the residents of both communities expressed concerns over both procedural and distributive justice. Injustice was perceived in the lack of information about siting the second repository and in the decision-making process, which lacked the voice and views of the communities' residents. In the case of the Puspokszilagy facility and the Ofalu attempt in Hungary, community members were highly interested in compensations and their distribution between the beneficiaries. However, there were protests due to a lack of justice in the general rule of distribution [6].

Trust is also a key issue in facility siting issues, which are always loaded with many risk components, such as social, environmental, and economic. Besides companies and industry, the government is a common object in which community members place their trust. Flynn et al. [16] found a connection between the level of trust in the government and the perception of risks and benefits of nuclear power; if trust in one's government is lower, then there is a higher perception of risks and a lower perception of benefits. Vainio et al. [17] state that it is important to take into consideration different kinds of trust, as they influence the acceptance of nuclear power, such as trust in ministries, nuclear energy companies and research institutes. Based on a national survey conducted in Finland, Vainio et al. [17] show that a higher level of trust in NGOs results in a decreased belief in nuclear power as a solution for climate change, whereas a higher level of trust in government and research institutes results in increased belief in nuclear power. Wang and Kim [18] analysed factors which affect the acceptance of nuclear power based on survey data from the Eurobarometer in 2009. The results show that the perceived benefits, trust and knowledge increase nuclear power acceptance. Regression analysis showed that the perceived benefits had a positive impact on acceptance in 26 out of 27 countries, trust had statistical significance in 23 countries, but knowledge had an effect in only eight countries. Another study, conducted by Litmanen et al. [19], regarding the project of building a final disposal repository in Finland showed interesting results concerning trust. Elements of trust were measured through two dimensions: trust in the company in charge and trust in the authorities regarding the risk assessment of the final disposal project. Results show that the residents' trust in the authorities was lower than their trust in the company [19]. Generally speaking, when a community trusts the guiding process by an organisation, there is a lower perception of risk and a higher level of acceptance. However, trust is not always a direct indicator that people will accept a siting proposal [20]. Community members can have trust in the process and industry, but they can also have other reasons for rejecting the project.

Although trust does not directly imply agreement with the project, it is hard to expect that there would be any local acceptance of a siting proposal without at least a sense of basic trust in the process of RAW management. This axiom serves as the foundation for the concept of the social licence to operate (SLO), developed by Joyce and Thomson [21] and Boutilier and Thomson [22], and later applied in the realm of RAW management by Lehtonen et al. [23]. In this adaptation, economic and socio–political legitimacy are again the key elements which build community trust, in this case in institutions responsible for RAW management. Optimisation of that process is indicated as the most useful approach toward the acquisition of SLO for establishing a RAW facility. In these terms, legitimacy refers to the acceptance of the project by the community, and it is mostly focused on justice. To gain acceptance, there must be economic legitimacy which is related to distributive justice. However, to win true approval, there must be socio–political legitimacy which includes procedural justice. In this case, the organisation must show a willingness to protect the community's life from the project's harmful impacts. This refers simultaneously to social, cultural, and environmental aspects of the community.

All these are essential for building trust, which serves as a bridge between, on the one hand, the legitimacy and, on the other, the acceptance, approval or identification with the project. Lehtonen et al. [23] conceptually differentiated between two levels of trust: interactional and institutionalised. The aspect of interactional trust implies confidence that the other side provides true information and is responsive to community concerns, which can be seen throughout envisioning new development goals with the community. Institutionalised or complete trust is a true indicator of the existence of SLO. In this case, interests from both sides are taken into consideration and citizens start to identify themselves psychologically with the interests and values of the organisation. They see the project as their own and start implementing their activities as part of the project. This dimension of trust develops as a result of a long-term interaction between the community, its environment and stakeholders [23]. While interactional trust can lead to a peaceful coexistence, only institutionalised trust can provide solid ground for all stakeholders to live long and prosperous lives.

As we have seen in previous studies, as well as in the SLO concept, there should be a connection between community trust-building and perception of justice of the project rationale regarding both its procedural and distributive components. Having that in mind, we were interested in how this mechanism works in a situation where the decision-making process for the site selection is still ongoing. More precisely, this paper aims to analyse relations between community trust in the state authorities in charge of the implementation of the national RAW programme and the perceived procedural and distributive justice related to the decision concerning the establishment of a centre for RAW management in Croatia.

3 STUDY AREA

In Croatia, RAW management has been carried out for many years. The existing RAW is temporarily stored at two locations in the city of Zagreb, and this is mainly material produced through the activities of medical institutions and scientific research laboratories. Given that the existing storage resources have been filled to capacity, and since Croatia due to the co-ownership of the Krško Nuclear Power Plant has an obligation to promptly take care of large quantities of low and intermediate-level RAW generated by the power plant so far, the state has intensified preparations for the establishment of a national RAW management centre [24].

Concerning the current legal framework regarding RAW management in Croatia, the Act on Radiological and Nuclear Safety was revised in the February of 2022 and is now in full harmony with the EU requirements for safe long-term management of RAW. This revision clearly defines the ultimate responsibility of the state for the safety of spent fuel and RAW management and provides normative changes required for implementation of the national programme for RAW management. According to the Act, the Fund for financing the decommissioning of the Krško NPP and the disposal of Krško NPP radioactive waste and spent nuclear fuel as a specialized institution is responsible for establishing the RAW Management Centre while the Ministry of the Interior as a regulatory body for radiological and nuclear safety is responsible for issuing an operation license. The RAW Management Centre is planned for storage of solid low and intermediate-level RAW only and is scheduled for commissioning in 2023. Current strategic documents for the spatial development of Croatia and the national programme for RAW management envisage the establishment of such a centre in the municipality of Dvor on the site of the former military logistics complex Čerkezovac.

The municipality of Dvor is located in the Sisak-Moslavina County on the border with Bosnia and Herzegovina, about 100 km south of Zagreb. According to the 2021 census, Dvor has a population of 3,044, which means that it has lost almost 80% of its inhabitants in the last three decades. In addition to demographic problems, the municipality of Dvor is characterised by a significant deficit in socio–economic development compared to the rest of Croatia. According to the latest calculation of the development index from 2018, Dvor is on the 548th place out of a total of 556 evaluated local self-government units in Croatia [25]. These indicators suggest that the living conditions which prevail in Dvor are quite unfavourable, and considering the broader picture, Opačić [26] estimates that this municipality falls into the category of one of the so-called disordered communities. These are communities which have undergone major and dramatic social changes in a certain period, which possess some resources, but not enough collective and institutional participants and the ability to use them for a stable and long-term economic growth and development. This is a consequence of a long term process. Even during the socialist period (1945–1990), Dvor was among the most underdeveloped municipalities in Croatia. The war and occupation in the first half of the 1990s only accelerated the collapse of the local economy and depopulation, and the post-war reconstruction was clearly either not ambitious enough or not successful in stopping these negative processes.

As mentioned in the introduction, the project of establishing a RAW management centre in Dvor is in its preparation phase. Collecting the necessary permits also includes the preparation of a feasibility study and an assessment of the impact of the project on environment. The basic principles according to which the process of establishing a RAW management centre takes place are defined by national legislation, as well as relevant recommendations of international institutions [25]. In addition to ensuring high technical standards, they include the application of an inclusive and participatory approach in the decision-making process. The purpose of the latter is to encourage building mutual trust and the establishment of a partnership with the local community. Given the unfavourable socio–economic situation in Dvor, the starting assumption is that the level of community trust in state authorities responsible for RAW management will be equally influenced by the perception of the procedural validity of the final decision and the belief that this arrangement can provide additional incentives to revitalise the local community.

4 STUDY DESIGN

The data for this study were collected through a survey conducted in the autumn of 2020 on a probabilistic stratified sample of adult residents in the municipality of Dvor. The stratification covered the following two elements: the size of the residential area and the distance of the settlement from the location designated for the future RAW storage facility. Based on these two criteria, a total of eight different survey zones were defined and the number of respondents from each of them was proportional to the share of the zone in the total population. The survey was conducted face-to-face at the respondents' residential address. The sample included a total of 424 respondents with a response rate of 52%.

The testing of the basic hypothesis was done using a hierarchical regression analysis. The gradual introduction of variables into the model enabled a comparison of the relative contribution of individual predictor sets or blocks, and we had a total of three blocks in this analysis. In the first block, we introduced the control variables, and then, in separate blocks, the perception of procedural justice, followed by the perception of distributive justice.

With the variables used in the model involving three or more items, a reduction was performed based on the results of a principal component analysis. Using the regression method, we generated factor points for the obtained factor solutions, so the created variables were used in further analysis. The instruments used in the model are presented below with the basic results of the principal component analysis, while the results of the descriptive analysis are presented in the next section.

The criterion variable in the analysis is the community trust in the state authorities in charge of RAW management. The instrument used to measure this type of trust was prepared based on Lehtonen et al.'s [23] theoretical framework which distinguishes between interactional and institutionalised trust. We can see that the key moment in interactional trust is confidence, and the instrument includes items which question the expectations of the local public that state authorities in charge of RAW management will fulfil their obligations and inform them about all events which may affect them. In the case of institutionalised trust, the emphasis is on the development of partnerships, so in the instrument we included the items which question the respect of local interests and the similarity of visions for the future development of the municipality. The variables included in our instrument are taken from Boutilier's [27] instrument for measuring the social licence to operate, but are adapted to specific circumstances in Croatia, i.e. that this is about establishing an infrastructure for RAW management and that the project is still in preparation. In total, our instrument contained four items, and the degree of agreement with them was measured on a 5-point Likert scale. The results collected from an instrument set in this way were subjected to a principal component analysis, which, using the Kaiser–Guttman rule, indicated a one-factor structure of the items used, explaining 60% of the total variance with Cronbach's α of 0.77.

In the analysis, we measured the effects of perception of distributive and procedural justice separately. The perception of distributive justice was captured by two elements. One was related to the expected personal benefits; we asked the respondents to rate on a 5-point scale whether the establishment of the centre will have a positive or a negative impact on their income, property value and personal life perspective. The second element affecting perception of distributive justice was related to communal benefits, i.e. on a 5-point scale we asked the participants about the level of agreement with four statements which expressed expectations that the project would improve living conditions and overall economic prosperity of the Dvor municipality. The principal component analysis was performed separately for each of the two elements, and with the application of the Kaiser–Guttman rule, a monofactorial structure was determined in both cases. The obtained factor solution explained 71% of the total variance in the perception of personal benefits (Cronbach's $\alpha = 0.80$), and 56% of the total variance in the perception of communal benefits (Cronbach's $\alpha = 0.73$).

In analysing the perception of procedural justice, we started from the premise that the basic purpose of an administrative procedure is to remove any doubts about the reasonability and validity of the decision on how and where to establish the infrastructure for RAW management. In that context, we focused on two aspects relevant to ensuring socio–political legitimacy. The first aspect refers to some form of international verification of the process of establishing such infrastructure, and we asked respondents about the extent to which they agreed with the statement that research on the suitability of the preferred location for the centre will be conducted in accordance with high international standards and EU directives. The second aspect refers to the role of the local government in the decision-making process, so we asked respondents about the level of agreement with the statement that cooperation between the municipality of Dvor and the state institutions responsible for RAW management is the best way to protect the local community's interests.

Furthermore, the model includes a set of control variables, and in the analysis of the relationship between criterion and predictor variables, in addition to basic socio–demographic characteristics (sex, age and education), we also isolated the effect of the distance of the place of residence from the planned infrastructure. The results of the descriptive analysis for the above control variables are shown in Table 1.

Table 1: The respondents' socio–demographic characteristics.

Sex	Male	46.2%
	Female	53.8%
Age	18–34	12.5%
	35–64	46.9%
	65+	40.6%
Educational attainment	Primary school or less	23.5%
	Secondary school	63.5%
	Higher education	13.0%
Distance from the planned RAW storage facility	< 5 km	20.0%
	5–10 km	37.5%
	> 10 km	42.5%

5 RESULTS

While analysing the collected data we first tried to determine the level of trust the local community has in the state authorities responsible for RAW management. The results shown in Table 2 reveal that the vast majority of respondents disagree with all of the claims offered. Specifically, 67% of respondents do not think that state authorities responsible for RAW management will fulfil their obligations, 71% do not think that the public will receive relevant information, 75% believe that the interests of the local community will not be taken into account, and 77% believe that the local community and state authorities responsible for RAW management do not share a similar vision of the future of this area. Thus, taken as a whole, these results clearly indicate that the majority of respondents do not actually see the willingness of the relevant state institutions to build partnerships with the local community.

Table 2: Results on statements related to community trust in state authorities in charge of RAW management.

State authorities in charge of RAW management:	Strongly or somewhat disagree	Neither agree, nor disagree	Strongly or somewhat agree
…will fully meet its commitments to our community.	66.9%	17.6%	15.5%
…will inform us about things that could affect our community.	71.3%	17.9%	10.8%
… are concerned about the interests of our community when making their decisions.	74.9%	14.8%	10.3%
…and our community have a similar vision for the future of this region.	76.6%	11.9%	11.4%

Furthermore, we focused on the perception of procedural justice by observing two items. One item referred to the importance of the local authorities' active participation in finding solutions that would be acceptable to the local community, and there were some divisions among the respondents. The results show there is a similar share of those who agree and those who disagree with the statement that cooperation between the municipality of Dvor and the state authorities responsible for RAW management is the best way to protect the interests of the local people (Table 3). The share of respondents in both groups was about 38%. When it comes to assessing the credibility of the site selection process, only 22% of respondents are convinced that it would be carried out in accordance with high international standards and EU directives, while 50% of respondents believe that this would not be the case. Let us also mention that in both claims a significant number of respondents were undecided, their share was 23% and 27% respectively.

Table 3: Results concerning statements related to the perception of procedural justice.

	Strongly or somewhat disagree	Neither agree nor disagree	Strongly or somewhat agree
Cooperative engagement of local authorities is the best strategy to protect local interests.	38.6%	23.3%	38.1%
Final decision about the location of the RAW storage facility will respect international standards and EU directives.	50.4%	27.1%	22.4%

Shifting the focus to the perception of distributive justice reveals that only few people expect to be able to personally benefit from the siting of a RAW management centre in their municipality. Even still, the results given in Table 4 at the same time show that not everyone thinks they would feel the negative consequences. The least expected negative impact is a reduced income in their household; about 45% of respondents see a possible danger there. With the other two items, the share of respondents who feel threatened is somewhat higher. About 70% of respondents expect negative consequences regarding the value of their property, and 73% believe that there will be negative consequences on their life prospects. In other words, although most do not expect an immediate dramatic collapse in standards, most respondents are inclined to believe that the arrival of RAW management infrastructure in their municipality will produce long-term negative consequences for their standard of living.

Table 4: Results concerning the statements related to the perception of personal benefits.

How would the establishment of the infrastructure for RAW management in Dvor affect the following aspects of your life:	Very or somewhat negative	Neither positive nor negative	Very or somewhat positive
Your household's income	44.8%	49.9%	5.4%
Value of your property	70.1%	25.8%	4.1%
Your personal life perspective	72.6%	24.0%	3.3%

An analysis of the perception of distributive justice at a communal level reveals that most respondents do not even expect the wider community to benefit from siting the RAW management infrastructure in their municipality (Table 5). Only few predict a positive effect on the overall local economy; raising the average living standard of the local population and accelerating the local economy is expected by only 10% and 9% of respondents respectively. However, there are slightly more optimistic respondents when assessing the improvement of infrastructure; 18% think that there will be an improvement in the supply of water, electricity and the Internet, and 32% expect the improvement of the transport infrastructure.

Table 5: Results concerning statements related to the perception of communal benefits.

	Strongly or somewhat disagree	Neither agree nor disagree	Strongly or somewhat agree
The local community infrastructure will be improved.	57.0%	24.5%	18.4%
Transport infrastructure will be improved.	49.4%	18.7%	31.9%
The standard of living of the local population will be increased.	78.8%	11.2%	10.0%
Local economic development will be enhanced.	79.5%	11.6%	9.0%

In the next step, using hierarchical linear regression analysis, we investigated the connection between community trust in the state authorities responsible for RAW management and the level of perceived procedural and distributive justice. Prior to the regression analysis, the correlation matrix review (Table 6) showed that there is no strong correlation between the predictor and control variables, which is a sign that the problem of multicollinearity was not evident in the model.

Table 6: Correlation matrix of variables used in the regression model.

	2	3	4	5	6	7	8	9
1. Community trust in state authorities in charge for RAW management	–.001	–.040	0	.119*	.324**	.500**	.285**	.444**
2. Sex	1	.035	–.254**	.018	–.033	.040	–.01	.061
3. Age		1	–.211**	.039	.063	–.025	–.026	–.145**
4. Educational attainment			1	–.144**	–.071	.029	.095	.095
5. Distance from the planned RAW storage facility				1	.168**	.186**	0	.174**
6. Procedural justice: cooperative engagement of local authorities is the best strategy to protect local interests					1	.387**	.145**	.251**
7. Procedural justice: decision about the location of the RAW storage facility will respect international standards and EU directives						1	.390**	.362**
8. Distributive justice: expected personal benefits							1	.225**
9. Distributive justice: expected communal benefits								1

* $p < 0.05$; ** $p < 0.01$.

The results shown in Table 7 reveal that the created regression model explains a total of 31% of the variance of community trust in the state authorities in charge for RAW management. The analysis of the model showed that socio–demographic variables do not significantly contribute to the understanding of changes in the criterion variable. The four control variables introduced in the first block explain only 2% of the variance, and only the distance of the place of residence from the planned location is statistically significantly positively related to the criterion variable. In the second block, two variables related to the perception of procedural justice were introduced. The percentage of the explained variance after the introduction of the second block was increased by 24 percentage points, and for both introduced variables we found a statistically significant correlation with the criterion variable. Those who believe that cooperation between the municipality and the state authorities in charge for RAW management is the best way to protect local interests and those who believe that the site decision will be made in accordance with high international standards and EU directives also show a higher level of trust in the state's good intentions that it actively supports the development of the local community by respecting its interests and needs. In this block, there was also a minor modification of the contribution of previously introduced variables to the change in the criterion variable. More precisely, after controlling for the effect of the perception of procedural justice, the distance of the planned infrastructure for RAW management from the place of residence is no longer related to the level of community trust.

Table 7: Regression results predicting community trust in state authorities in charge of the RAW management concerning the perception of procedural and distributive justice.

Predictors	Step 1 β	Step 2 β	Step 3 β
Socio–demographics			
Sex	0.004	0.004	–0.021
Age	–0.042	–0.030	–0.003
Educational attainment	0.071	0.055	0.009
Distance from the planned RAW storage facility	0.172**	0.060	0.034
Perception of procedural justice			
Cooperative engagement of local authorities is the best strategy to protect local interests		0.168**	0.134*
Final decision about the location of the RAW storage facility will respect international standards and EU directives		0.407**	0.309**
Perception of distributive justice			
Expected personal benefits			0.095
Expected communal benefits			0.235**
Explained variance			
Adjusted R-square	0.022	0.258	0.311
F	2.823*	19.816**	19.271**
Δ Adjusted R-square	–	0.236	0.053

* $p < 0.05$; ** $p < 0.01$.

In the third and final block, we introduced two variables related to the perception of distributive justice. They increased the proportion of the explained variance by an additional 5 percentage points. However, just one of the two introduced variables in the third block was significantly positively related to the criterion variable; only the expectation of higher communal benefits was associated with a higher level of trust in the state authorities in charge for RAW management. In the final model, there was no change in the contribution of other previously introduced variables. Therefore, after controlling for other covariates included in the final model, both variables related to the perception of procedural justice and the variable related to the expected communal benefits were positively related to the level of community trust in the relevant state institutions. In doing so, the block with the perception of the distributive justice contributed relatively less to the total amount of explained variance than the block with the perception of procedural justice.

6 DISCUSSION AND CONCLUSION

The analysis presented in this study was an attempt to contribute to the understanding the local community's attitude towards the plan for the establishment of an infrastructure for the management of low and intermediate level RAW in its local area based on empirical insights. We mentioned that modern approaches to solving the problem of RAW management start from the fact that it is not exclusively a technical problem, but largely depends on social perception. In this context, building partnerships with the local community when creating one such infrastructure is evidently one of the most important prerequisites for successful project implementation. Important components of the development of a partnership are certainly the belief in the reasonability and validity of the decision itself, i.e. the social perception of the legitimacy of the decision and trust in the actors who make and implement decisions [8]. It is the latter which forms the backbone of the concept of social licence to operate [23]. According to this model, the relationship between the local public and the RAW management infrastructure can be graded in the range from rejection, through acceptance and tolerance, to approval, all the way to identification, and the outcome depends on the level of established mutual trust. The key to building trust lies in the way the local public perceives the economic and socio–political legitimacy of the project itself, i.e. how it perceives the elements of procedural and distributive justice. Our analysis focused on that very moment. In this paper, we investigated the relationship between community trust in the state authorities responsible for RAW management and, on the one hand, the perception of assurance that the procedure would take into account the protection of local interests, as well as protection of environment and health, and, on the other hand, the perception of possible benefits from the project, both for them personally and for the local community as a whole. The data analysed in the paper were collected by a survey conducted in the autumn of 2020 on the sample of the inhabitants of the municipality of Dvor, whose local area is a preferred location for the construction of a national centre for RAW management.

The conducted analysis provided several useful insights. Firstly, the respondents expressed a low level of trust in the intention of the state authorities responsible for the management of RAW to build partnerships with the local community. For example, three out of four respondents doubt their readiness to take into account the local community's interests and build a common vision of local development, and two thirds even doubt that the promises and commitments to the local community would be fulfilled. The low level of trust in the state authorities responsible for RAW management is not specific to Croatia, but is for various reasons also present in countries with much more developed democratic institutions [28]. The reasons for the low level of trust observed in our study should be sought primarily in the deficit of direct communication between the local community and the state authorities responsible for RAW management. Admittedly, there were already opportunities for the local public to express their opinions at some stages of the decision-making process, for example, when drafting the national program for implementing the RAW management strategy. However, the impact of their views on the final version of the document was negligible [24]. In addition, what Blowers [29] characterises as *the persistence of the periphery* seems to be important for understanding the relationship between the local community in Dvor and the state authorities responsible for RAW management. Moreover, Dvor is not only geographically in a peripheral position, but also encounters the problem of insufficient or inadequate basic infrastructure, a large number of economically deprived population dependent on social transfers with the collapse of key elements of the community (institutions and local economy), exhibiting the need to restore community identity in the new circumstances after the dramatic events of the Croatian War of Independence. All this contributes to the fact that a part of the population shares a certain feeling of stigmatisation and powerlessness in the community, as it seems that everything is happening against their image of development [26]. In the context of such significant social and economic deprivation, the imposition of infrastructure for the management of RAW in the Dvor area can easily be seen from the local perspective as another proof of the state's neglect of its local community.

Secondly, the results of the regression analysis confirmed the initial hypothesis about the relationship between trust in the state authorities responsible for RAW management and the perceptions of procedural and distributive justice. More specifically, the analysis showed that besides the expectation that the RAW management infrastructure at Dvor would enhance the local standard of living, the belief in improving the credibility of procedures in ensuring adequate protection of environment and health was also associated with a higher level of confidence in the state's good will to cooperate with the local community. However, the perception of procedural justice had a slightly stronger effect than the perception of distributive justice on the change of the criterion variable. Moreover, in the presented model, the expectation of possible personal benefits from siting the infrastructure in Dvor did not prove to be relevant for the level of the expressed trust in the state authorities responsible for RAW management. These are consistent with the findings of a study by Krütli et al. [13], that the perception of procedural justice in the process of establishing RAW management infrastructure exceeds the importance of the perception of distributive justice. Di Nucci and Brunnengräber [8] reached a similar conclusion while analysing several examples of RAW management infrastructure placement, noting that the local community still expects some involvement in the decision-making process before discussing rent and benefits, as well as strengthening its capacity to understand the challenges arising from coexistence with such infrastructure.

Finally, none of the control variables used in the model showed a statistically significant effect on community trust in the government authorities responsible for RAW management. In other words, there is a local consensus on the importance of the perception of justice for the establishment of community trust, and here there is no division according to sex, age or education, nor even according to whether people live close to or somewhat further from the preferred location.

In conclusion, our findings confirm the operation of a basic mechanism that is at the heart of the concept of social licence to operate: stronger perception of legitimacy are reflected to higher levels of community trust. Regarding certain limitations of the present study, it should be noted that we have dealt with only one segment in the complex process of acquisition of a social licence to operate. Lehtonen et al. [23] draw attention to the importance of contextual factors in this process, primarily to the previous experience in RAW management and the level of general trust in the state. In this sense, one should be careful in generalising our results, because previous research shows that the Croatian public is largely uninformed about this issue [24] and has relatively low trust in institutions [30]. In addition to testing the functionality of this model in another context, future research could test a more complex regression model. For example, it would be useful to see what effect the perception of risk has on the relationship between perceived legitimacy and trust in the RAW management institutions, which has already been reported as a major determinant of (un)acceptability of nuclear energy [31] as well as of new infrastructure for RAW management [32].

REFERENCES

[1] Marshall, A., The social and ethical aspects of nuclear waste. *Electronic Green Journal*, **21**(1), 2005. DOI: 0.5070/G312110587.

[2] Ramana, M.V., Technical and social problems of nuclear waste. *WIREs Energy and Environment*, 7(4), e289, 2018. DOI: 10.1002/wene.289.

[3] Waldman, S.M., Taking risk seriously: Discourses and worldviews in a nuclear waste controversy, Dissertation, Carleton University, Ottawa, 2017. DOI: 10.22215/etd/2017-12185.

[4] Bergmans, A., Sundqvist, G., Kos, D. & Simmons, P., The participatory turn in radioactive waste management: Deliberation and the social-technical divide. *Journal of Risk Research,* **18**(3), pp. 347–363, 2015. DOI: 10.1080/13669877.2014.971335.

[5] Tomczak, W., Socioeconomic aspects in the development and operation of the national radioactive waste repository: Rozan. *Proceedings of the Workshop Low and Intermediate Level Waste Repositories: Socioeconomic Aspects and Public Involvement,* pp. 83–87, 2007.

[6] Vari, A. & Frerencz, Z., Radioactivewaste management in Hungary: Changing approaches and conflicts. *Journal of Environmental Assessment Policy and Management,* **9**(2), pp. 185–209, 2007.

[7] Perry, J., Trust in public institutions: Trends and implications for economic security. *UN Department of Economic and Social Affairs: Policy Brief,* **108**, pp. 1–4, 2021.

[8] Di Nucci, M.R. & Brunnengräber, A., In whose backyard? The wicked problem of siting nuclear waste repositories. *European Policy Analysis*, **3**(2), pp. 295–323, 2017. DOI: 10.1002/EPA2.1028.

[9] Cotton, M., *Nuclear Waste Politics: An Incrementalist Perspective*, Routledge: New York, 2017.

[10] Di Nucci, M.R. & Brunnengräber, A., Making nuclear waste problems governable: Conflicts, participation and acceptability – The German case. *Conflicts, Participation and Acceptability in Nuclear Waste Governance: An International Comparison*, Vol. III, eds A. Brunnengräber & M.R. Di Nucci, Springer: Wiesbaden, pp. 3–21, 2019.

[11] Wolsink, M., Planning of renewables schemes: Deliberative and fair decision-making on landscape issues instead of reproachful accusations of non-cooperation. *Energy Policy*, **35**(5), pp. 2692–2704, 2007.

[12] Wüstenhagen, R., Wolsink, M. & Bürer, M.J., Social acceptance of renewable energy innovation: An introduction to the concept. *Energy Policy*, **35**(5), pp. 2683–2691, 2007.

[13] Krütli, P., Stauffacher, M., Pedolin, D., Moser, C. & Scholz, R.W., The process matters: Fairness in repository siting for nuclear waste. *Social Justice Research*, **25**(1), 79–101, 2012. DOI: 10.1007/s11211-012-0147-x.

[14] Krütli, P., Törnblom, K., Wallimann-Helmer, I. & Stauffacher, M., Distributive versus procedural justice in nuclear waste repository siting. *The Ethics of Nuclear Energy: Risk, Justice and Democracy in the Post Fukushima Era*, eds B. Taebi & S. Roeser, Cambridge University Press: Cambridge, pp. 119–140, 2015.

[15] Vilhunen, T., Kojo, M., Litmanen, T. & Taebi, B., Perceptions of justice influencing community acceptance of spent nuclear fuel disposal: A case study in two Finnish nuclear communities. *Journal of Risk Research*, 2019. DOI: 10.1080/13669877.2019.1569094.

[16] Flynn, J., Burns, W., Mertz, C.K. & Slovic, P., Trust as a determinant of opposition to a high-level radioactive waste repository: Analysis of a structural model. *Risk Analysis,* **12**(3), pp. 417–430, 1992.

[17] Vainio, A., Paloniemi, R. & Varho, V., Weighing the risks of nuclear energy and climate change: Trust in different information sources, perceived risks, and willingness to pay for alternatives to nuclear power. *Risk Analysis*, **37**(3), pp. 557–569, 2017.

[18] Wang, J. & Kim, S., Comparative analysis of public attitudes toward nuclear power energy across 27 European countries by applying the multilevel model. *Sustainability,* **10**(5), pp. 2–21, 2018.

[19] Litmanen, T., Kojo, M. & Kari, M., The rationality of acceptance in a nuclear community: Analysing residents' opinions on the expansion of the SNF repository in the municipality of Eurajoki, Finland. *International Journal of Nuclear Governance, Economy and Ecology*, **3**(1), pp. 42–58, 2010.

[20] Sjöberg, L. & Drottz-Sjöberg, B.-M., Fairness, risk and risk tolerance in the siting of a nuclear waste repository. *Journal of Risk Research*, **4**(1), pp. 75–101, 2001. DOI: 10.1080/136698701456040.

[21] Joyce, S. & Thomson, I., Earning a social licence to operate: social acceptability and resource development in Latin America. *Canadian Mining Metallurgical Bulletin*, **93**, pp. 49–53, 2000.

[22] Boutilier, R. & Thomson, I., Modelling and measuring the social licence to operate: Fruits of a dialogue between theory and practice, 2011. www.socialicense.com. Accessed on: 10 Feb. 2022.

[23] Lehtonen, M., Kojo, M., Jartti, T., Litmanen, T. & Kari, M., The roles of the state and social licence to operate? Lessons from nuclear waste management in Finland, France, and Sweden. *Energy Research and Social Science*, **61**, 2020. DOI: 10.1016/J.ERSS.2019.101353.

[24] Trontl, K., Pevec, D., Jakić, I. & Matijević, M., Radioactive waste management in Croatia: Public opinion, legal framework, and policy. *Energy Policy*, **146**, 2020. DOI: 10.1016/j.enpol.2020.111804.

[25] Ministry of Regional Development and EU Funds, Vrijednosti indeksa razvijenosti i pokazatelja za izračun indeksa razvijenosti jedinice lokalne samouprave. www.razvoj.gov.hr/. Accessed on: 10 Feb. 2022.

[26] Opačić, A., Three types of underdeveloped communities in Croatia: Latent, collapsed, and disorganized communities. *Community Development*, **51**(5), pp. 575–592, 2020. DOI: 10.1080/15575330.2020.1807580.

[27] Boutilier, R., A measure of the social license to operate for infrastructure and extractive projects, 2017. www.socialicense.com. Accessed on: 10 Feb. 2022.

[28] Seidl, R., Krütli, P., Moser, C. & Stauffacher, M., Values in the siting of contested infrastructure: The case of repositories for nuclear waste. *Journal of Integrative Environmental Sciences*, **10**(2), pp. 107–125, 2013. DOI: 10.1080/1943815X.2013.824486.

[29] Blowers, A., *The Legacy of Nuclear Power,* Routledge: Abingdon, 2017.

[30] Nikodem, K. & Črpić, G., O (ne)održivosti veza između povjerenja i demokracije: Analiza povjerenja u institucije u Hrvatskoj i u Europi. *Vrednote u Hrvatskoj i u Europi,* eds J. Baloban, K. Nikodem & S. Zrinščak, Kršćanska sadašnjost: Zagreb, pp. 259–307, 2014.

[31] de Groot, J.I.M., Schweiger, E. & Schubert, I., Social influence, risk and benefit perceptions, and the acceptability of risky energy technologies: An explanatory model of nuclear power versus shale gas. *Risk Analysis*, **40**(6), pp. 1226–1243, 2020. DOI: 10.1111/risa.13457.
[32] Flynn, J., Slovic, P. & Mertz, C.K., Decidedly different: Expert and public views of risks from a radioactive waste repository. *Risk Analysis*, **13**(6), pp. 643–648, 1993. DOI: 10.1111/j.1539-6924.1993.tb01326.x.

NETWORK STRUCTURES IN HAZARDOUS WASTE MANAGEMENT SYSTEMS: ADVANCES, CHARACTERISTICS AND CHALLENGES

PAULA A. CABRALES NAVARRO, JAVIER E. ARIAS OSORIO & JULIO C. CAMACHO PINTO
School of Industrial and Business Studies, Industrial University of Santander, Colombia

ABSTRACT

In recent years, progressive industrial and technological development has increased the amount of hazardous waste generated on the planet. The improper management of these wastes adversely affects human health and the environment due to their corrosive, reactive, ignitable, and toxic characteristics. Consequently, the activities associated with the collection, transport, treatment, recycling, and disposal of hazardous waste increase the related risks producing crucial economic and social impacts. For this reason, hazardous waste management attracts researchers' attention who through the formulation of optimization models have proposed in the literature the hazardous waste location-routing problem (HWLRP). In this regard, this paper studies the evolution of network structures in HWLRP existing in the literature. The main focus of this paper is to provide a review of the most relevant frameworks for the design of a hazardous waste management system and to analyze advances, characteristics, and challenges for future research. This study may help to define assumptions, concepts, and some insights into the design of the subsequent location and routing models applied to hazardous waste management to minimize the economic, social, and environmental impact that these wastes generate.

Keywords: waste management systems, hazardous waste, network structures, location and routing.

1 INTRODUCTION

Hazardous wastes have characteristics such as corrosivity, reactivity, flammability, and toxicity [1]. These wastes are mainly generated by different sources, from large industries, such as chemical factories and oil refineries [2], and small businesses, homes, or hospitals [3]. Industrial and technological development worldwide in recent years, together with population and economic growth, have considerably increased the generation of hazardous waste, that is, there is a direct relationship between the level of development of the country and the quality of life of people concerning the amount of waste produced [4]. Taking into account its characteristics, hazardous waste represents a potential risk to human health and the environment, therefore, it requires strict control from its collection to its final disposal. For this reason, hazardous waste management systems (HWMS) have been created, which are responsible for the collection, distribution, treatment, reuse, and disposal of this waste efficiently and safely [5].

The design of a hazardous waste management system involves various strategic, tactical, and operational decisions, such as the location of recycling or treatment centers and disposal facilities, and the distribution of the routes that collect the waste from the generation nodes [6]. These decisions have been extensively studied in the literature through the application of optimization models in three categories: facility location problem (FLP), vehicle routing problem (VRP), and the combination of these models known as the hazardous waste location-routing problem (HWLRP), which is considered the problem of greatest interest to researchers because it jointly solves the most important decisions for the design of a hazardous waste management system. In these systems, the selection of waste collection routes depends considerably on the location of the treatment, recycling, and disposal centers. According to this, the location of the facilities affects the distribution of the routes for the

www.witpress.com, ISSN 1743-3541 (on-line)
doi:10.2495/WMEI220141

transport of waste, therefore, these decisions are connected, and considering the problem separately can generate an increase in costs and risks [7].

Unlike the traditional location and routing problem, which mainly seeks to minimize total costs, hazardous waste management models have been addressed with different objectives. The selection of the safest routes for waste collection is an important objective, but this minimization of risk can result in increased costs. Whereas, if the main decision is to reduce costs, the risk to society or the environment may be increased. For this reason, most of the existing models in the literature use multi-objective optimization to make the system more efficient and give a different solution alternative to the decision-maker. Cost minimization and risk minimization are the main objectives studied by researchers [3].

An important feature in the design of a hazardous waste management system is the network structure for the location of facilities and the transport of waste between these facilities. The evolution of location and routing models for hazardous waste over the years has presented different frameworks for the network structure with different characteristics, taking into account the need to bring the system closer to real-world conditions. The purpose of this article is to carry out an analysis of the evolution of the main frameworks presented in the literature for the hazardous waste location-routing problem, emphasizing the advances and characteristics of each one of them. In this way, a general framework of the system is proposed so that it serves as a reference for subsequent studies applied to hazardous waste management.

The rest of this article is organized as follows: Section 2 describes a literature review on the HWLRP, and Section 3 presents the advances in network structures for hazardous waste management systems and their characteristics. Section 4 presents the discussion, design of the framework, and challenges for future research, and finally, Section 5 describe the conclusions.

2 REVIEW OF HAZARDOUS WASTE LOCATION-ROUTING PROBLEM

The design of hazardous waste management systems has been mostly addressed in the literature as location and routing problems, therefore, this review presents the most relevant research on this topic. The first studies on the HWLRP began in the late 1980s when Zografos and Samara [8] designed for the first time a multi-objective hazardous waste management distribution network that simultaneously considered the minimization of site risk, transport risk, and travel time, for a single type of waste. Later, ReVelle and Jared [9] developed a model to minimize transportation costs and risk for the location of storage and transportation facilities in a problem applied to waste from the nuclear sector. That same year, List and Mirchandani [10] were the first to propose a model with different types of hazardous waste and multiple types of treatment technologies, considering the minimization of the total risk (transport risk and site risk), transport costs, and risk equity maximization. This last objective was also presented by Current and Ratick [11]. Giannikos [7] formulated a problem for the location of disposal facilities and the transportation of hazardous waste with four objectives: minimize costs, total risk, risk equity, and equitable distribution of the disutility caused by the operation of the treatment facilities. A new model is proposed by Nema and Gupta [12] who formulated waste–technology compatibility constraints, taking into account that hazardous waste cannot be treated together unless they are compatible with each other.

At the beginning of the new century, one of the most important studies was presented. Alumur and Kara [2] formulated a model for a real case in the central region of Anatolia, Turkey, to determine the location of treatment centers, disposal facilities, and the transportation of different types of waste according to compatibility constraints to minimize risks and costs. In the same year, Caballero et al. [13] addressed the location of incineration plants for the disposal of solid animal waste and designed routes to serve the generation nodes. Another outstanding study was presented by Samanlioglu [1] who proposed a new design for the network structure, which includes: waste generation nodes, treatment centers, recycling centers, disposal facilities, and the routes between them to minimize total costs, transportation risk, and site risk. Zhao and Verter [14] carried out an application to the used oil waste management, where the routes were formulated for the first time as a tour. Zhao et al. [15] incorporated into their model, the capacity and operation constraints, and risk tolerance measures between routes. An important contribution to the literature on hazardous waste was presented by Yilmaz et al. [5] who made a taxonomy of hazardous waste, establishing seven classifications in total with their respective treatment processes.

In 2018, Rabbani et al. [3] developed for the first time a model with a heterogeneous fleet of vehicles, considering waste–waste compatibility because hazardous waste should not be collected in the same vehicle. In addition to this, the design of the network structure considered transportation as a vehicle routing problem for collection at generation nodes, a condition that had not been previously studied. Zhao and Huang [16] addressed a gap not considered in the literature, since they considered a multi-period planning horizon for location decisions in their model, while Rabbani et al. [17] formulated a multi-objective model with multiple periods for vehicle routing and transportation decisions between facilities. Farrokhi-Asl et al. [18] considered environmental decisions as one of the study objectives of their model. Previous studies optimized costs and risks mainly. Later, Delfani et al. [19] proposed the concept of green location-routing problem (GLRP) applied to hazardous waste, with minimization of CO_2 emissions within its objectives.

Rabbani et al. [20] developed an extension of the models presented in previous studies, including a new optimization objective: workload balance and time windows for the vehicle routing problem. Finally, Tirkolaee et al. [21] addressed a location and routing model with time windows for medical waste management, as a problem that has increased due to the COVID-19 pandemic. The review carried out allows us to recognize the different hazardous waste location-routing models existing in the literature and their main characteristics as evidence of the importance of the research topic. The following section highlights the most important network structures for a hazardous waste management system that have been implemented to solve these models and their evolution according to the findings of the literature review carried out.

3 NETWORK STRUCTURES IN HAZARDOUS WASTE MANAGEMENT SYSTEMS

The planning of a hazardous waste management system begins with the design of the network structure that allows determining the required facilities and the flow of activities that are part of the system. To guarantee the efficient collection, distribution, treatment, reuse, and disposal of hazardous waste, the network structure must be made up of a set of facilities such as recycling centers, treatment centers, and disposal sites, and to define how the transporting process is carried out. Table 1 shows how the different existing frameworks in the literature for the design of network structures in hazardous waste management are made up.

As shown in Table 1, the first studies describe a structure made up only of waste generation nodes and disposal facilities where the waste is sent for disposal. Most studies focused on the collection of a single waste and a single treatment technology. However, in List and Mirchandani [10] these conditions were changed to model the collection of various types of waste. Later, other important facilities were included in the network structure, such as hazardous waste treatment centers with their different treatment technologies, and later, recycling centers, because part of this waste can be recovered both in generation nodes and in treatment centers.

Table 1: System framework on the hazardous waste network design.

Author	Year	Ref	System framework						
			O	G	R	T	D	W	Q
Zografos and Samara	1989	[8]		✓			✓	S	S
ReVelle and Jared	1991	[9]		✓			✓	S	S
List and Mirchandani	1991	[10]		✓			✓	M	M
Current and Ratick	1995	[11]		✓			✓	S	S
Giannikos	1998	[7]		✓			✓	S	S
Nema and Gupta	1999	[12]		✓		✓	✓	M	M
Alumur and Kara	2007	[2]		✓		✓	✓	M	M
Caballero et al.	2007	[13]		✓			✓	S	–
Samanlioglu	2013	[1]		✓	✓	✓	✓	M	M
Zhao and Verter	2015	[14]		✓			✓	S	–
Zhao et al.	2016	[15]		✓	✓	✓	✓	M	M
Yilmaz et al.	2017	[5]		✓	✓	✓	✓	M	M
Rabbani et al.	2018	[3]	✓	✓	✓	✓	✓	M	M
Zhao and Huang	2019	[16]		✓	✓	✓	✓	M	M
Rabbani et al.	2019	[17]	✓	✓	✓	✓	✓	M	M
Farrokhi-Asl et al.	2020	[18]	✓	✓	✓	✓	✓	M	M
Delfani et al.	2020	[19]		✓	✓	✓	✓	M	M
Rabbani et al.	2021	[20]	✓	✓	✓	✓	✓	M	M
Tirkolaee et al.	2021	[21]		✓			✓	S	–

Note: O = depot/origin, G = generation node, R = recycling center, T = treatment center, D = disposal facility, W = waste, Q = treatment technology, S = single waste or technology, M = multiple waste or technology.

Finally, recent studies model waste collection between generation nodes as a vehicle routing problem, which includes depots where vehicles are located. Next, the characteristics of the most relevant frameworks in the literature are described to recognize the advances over the years in the design of network structures for hazardous waste management.

3.1 Generation nodes and disposal facilities

In this network structure, two decisions must be resolved: the location of the disposal facilities and the transport of waste from the generation nodes to the previously located sites, as shown in Fig. 1. Each generation node sends the collected waste (a single type of waste) to a single disposal facility, that is, the transport flow is direct between the nodes (origin–destination). Studies with this framework were developed by Giannikos [7], Zografos and Samara [8], ReVelle and Jared [9], List and Mirchandani [11], and Caballero et al. [13].

Advantages

- Location and routing models with this network structure are less complex to solve.
- These studies are considered the point of departure for subsequent research on hazardous waste management.

Disadvantages

- Considering a single type of waste is far from the real conditions regarding hazardous waste.
- Recycling is not taken into account within hazardous waste management Desventajas.

Figure 1: G+D framework.

3.2 Generation nodes, treatment centers, and disposal facilities

In this network structure, initially proposed in Nema and Gupta [12] and improved and implemented in a real case study years later by Alumur and Kara [2], some nodes generate hazardous waste which can be recycled or sent to a treatment center for its processing. In the treatment centers, part of hazardous waste is recycled and the waste residue is produced, which is transported to the disposal facilities for its elimination (Fig. 2). In this hazardous waste management system, treatment centers and disposal facilities are located within a set of previously defined potential sites and the amount of waste that is transported between the selected facilities is determined.

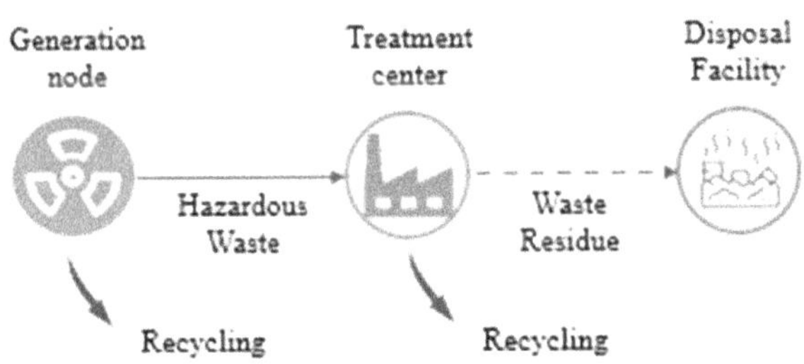

Figure 2: G+T+D framework.

Advantages

- The network structure considers different types of waste and different treatment technologies (one type of technology in each treatment center).
- The proposed models consider waste–technology compatibility constraints.
- Recycling is allowed both in the waste generation nodes and in the treatment centers.
- Mass balance constraints are considered for waste transportation between treatment centers and disposal facilities, that is, hazardous waste that cannot be treated or recycled is sent to final disposal facilities for elimination.

Disadvantages

- The system with this network structure does not take into account the location of recycling centers.

3.3 Generation nodes, treatment centers, recycling centers, and disposal facilities

The location of recycling centers in the system is included for the first time in Samanlioglu [1]. The design implemented by this author is considered one of the most complete available in the literature for hazardous waste management [18]. The network structure consists of some waste generation nodes and three types of facilities: treatment, recycling, and disposal. As shown in Fig. 3(a), after storing the different types of hazardous waste in the generation nodes, two activities are carried out: sending non-recyclable waste to a compatible treatment

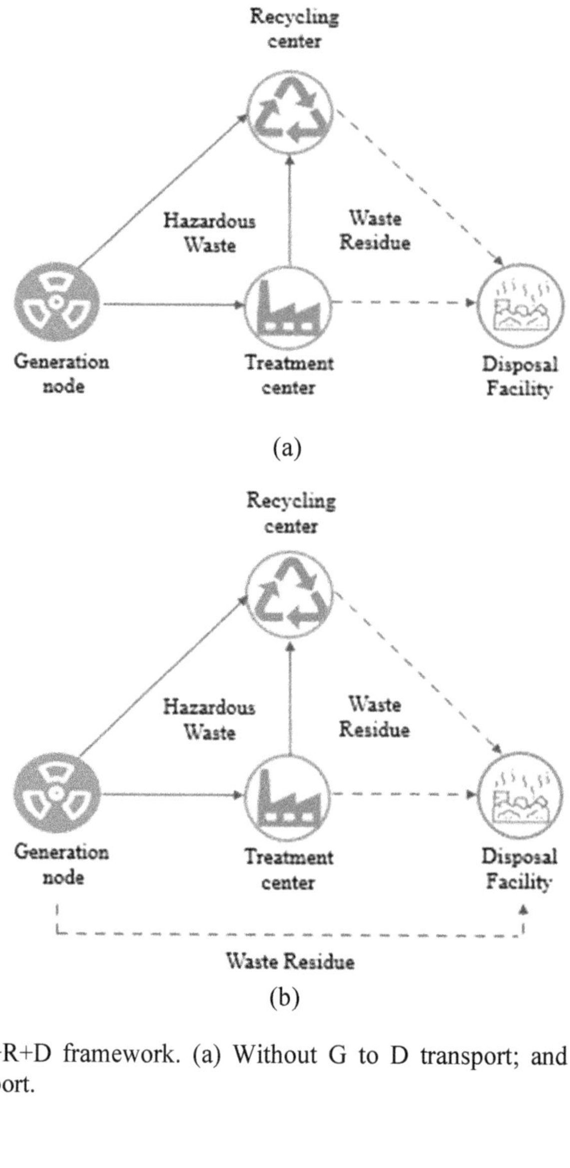

Figure 3: G+T+R+D framework. (a) Without G to D transport; and (b) With G to D transport.

center and sending recyclable waste to a recycling center. In a treatment center, after carrying out the transformation processes, waste is generated that is transported to a recycling center or sent to a disposal facility. In the same way, in the recycling centers, waste that cannot be reused is discarded and sent to the disposal facility. In this system, there is no direct flow of waste from the generation nodes to the disposal facilities.

Fig. 3(b) presents a network structure that allows waste transportation from the generation nodes to the disposal facilities. In this design proposed in Zhao et al. [15] and Zhao and Huang [16], hazardous waste is classified according to its generation source: treatable, recyclable, and disposable. In this way, disposable waste is sent directly for disposal in landfills.

Advantages

- In addition to the advantages associated with the structure presented in the previous section, this system takes into account the importance of recycling, therefore, it allows the location of recycling centers and the transport of waste between treatment centers to recycling centers and between recycling centers to disposal facilities.

Disadvantages

- The system does not consider the vehicle routing problem for waste collection between generation nodes. Although most of the previous studies define hazardous waste management as a location and routing problem, they do not explicitly deal with the classic problem proposed by Nagy and Salhi [22]. In these cases, routing is associated with waste transportation between facilities and generation nodes.

3.4 Depot, generation nodes, treatment centers, recycling centers, and disposal facilities

To address the disadvantage presented in the previous section, a network structure that considers the vehicle routing problem is presented in Rabbani et al. [3]. Although the facilities to be located are the same as those presented by Samanlioglu [1], for this structure exists a central depot where the vehicles are parked. The vehicles start their route at the depot, collect the hazardous wastes at the generation nodes, drop off their load at a recycling center or a treatment center compatible with the waste they transport, and finally return to the depot. In addition to this, the waste transport processes between facilities mentioned in the previous structures are carried out (Fig. 4).

Studies with this type of network structure have also been presented in Rabbani et al. [17], [20]. According to Farrokhi-Asl et al. [18] additionally allows the location of different depots within a set of potential sites.

Advantages

- As in the previous structure, the implementation of different types of waste, different treatment technologies, location of treatment centers, recycling, and disposal facilities are considered an advantage.
- Implementation of the vehicle routing problem for the collection of hazardous waste.
- Waste–waste compatibility constraints are formulated, taking into account that hazardous waste cannot be transported together.
- For this, a system is designed for the first time that proposes the use of a heterogeneous fleet of vehicles for the collection of the waste generated.

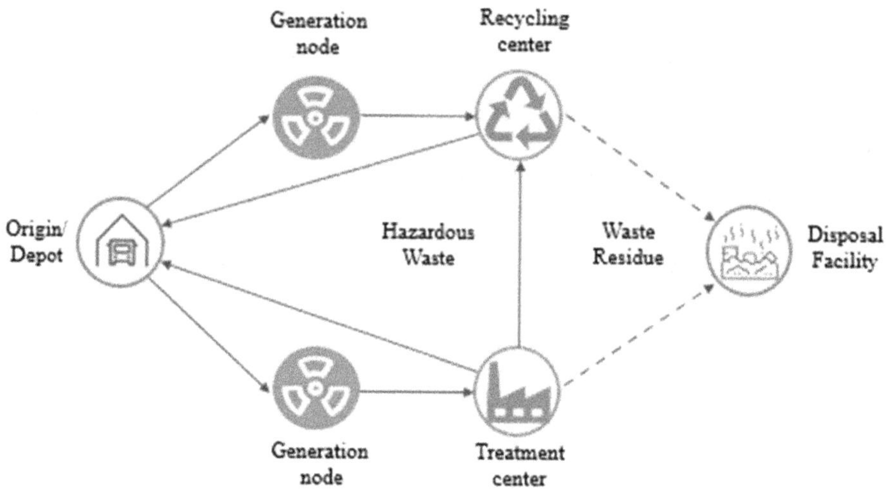

Figure 4: O+G+T+R+D framework.

Disadvantages

- The network structure increases the complexity of the hazardous waste management system, which represents the need to implement more robust solution methods for the multi-objective location and routing problem.

4 DISCUSSION: FRAMEWORK AND CHALLENGES

The evolution of network structures for the design of a hazardous waste management system allows us to recognize the importance of this research topic over the years. The planning of these systems implies the selection of different facilities, the collection of waste at the generation points, and the transport of waste residue between the selected facilities [12]. All of the above is an integrated process that can be solved by formulating multi-objective location and routing mathematical models. Studies developed in the literature have particular characteristics to adapt their models to conditions closer to real life. Although the first studies did not take into account any of these conditions, technological advances and the development of new solution strategies for highly complex mathematical models have allowed problems to be formulated that demand greater computational efforts but that contemplate a more complete design. Thus, an effective hazardous waste management system should achieve the following:

- Establishing the routes of the vehicles that are responsible for the collection of hazardous waste in each of the generation nodes.
- Determining the location of the system facilities: treatment centers, recycling centers, and disposal facilities.
- Determining the type of technology to be implemented in each localized treatment center.
- Establishing the transport process for recyclable waste and waste residue to be disposed of between the selected facilities.

4.1 General framework

According to the findings of the research carried out, this article presents a general framework that integrates all the previously mentioned decisions and serves as a point of departure for the design of a hazardous waste management system. This allows future research on the hazardous waste location-routing problem to take into account the characteristics of this framework for the network structure. The proposed framework is shown in Fig. 5. First, the generation nodes of hazardous waste are presented, which can be factories, hospitals, homes, and small businesses, among others. In each generation node, the waste is classified as recyclable waste, treatable waste, and disposable waste. Once the waste is stored at the source, a fleet of vehicles carries out the collection process. It must be taken into account that the dangerous conditions of this waste do not allow some of them to be transported together, therefore, a different type of vehicle is available for the collection of each classified waste. In this way the following activities are carried out:

1. Vehicles that collect recyclable hazardous waste send their load to a recycling center.
2. Vehicles that collect treatable hazardous waste send their load to a treatment center considering that each type of waste can only be treated by one type of compatible treatment technology. Hazardous waste treatment is generally carried out using chemical and thermal methods. Among the most used chemical processes are ion exchange, oxidation, and precipitation, while thermal processes are mainly applied through incineration [3].
3. Vehicles that collect disposable hazardous waste send their load directly to a disposal facility.

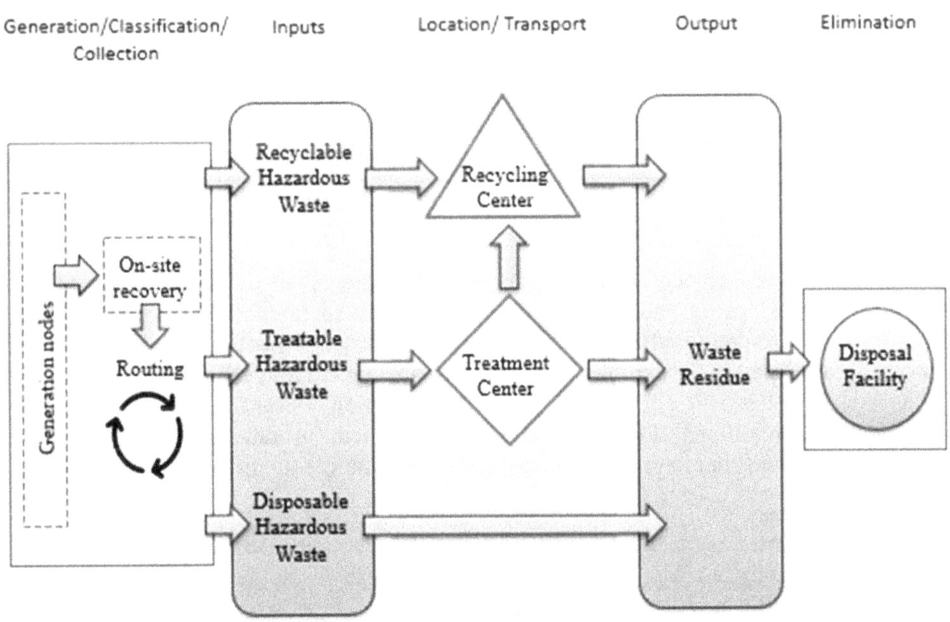

Figure 5: Hazardous waste management system framework.

Subsequently, in the treatment and recycling centers, transformation processes are carried out with the following activities:

1. In the recycling centers, a percentage of the waste is recovered while the waste residue is sent to a disposal facility.
2. In the treatment centers, a percentage of the waste becomes recyclable and is sent to a recycling center, and the waste residue is also sent to a disposal facility.

Table 2 presents some examples of the different hazardous waste according to their classification, taking into account the research carried out in Yilmaz et al. [5]. For treatable waste, waste compatible with chemical treatment and waste compatible with incineration are separated.

Table 2: Hazardous waste classification and examples.

Recyclable waste	Treatable waste		Disposable waste
	Chemical treatment	**Incineration**	
Wood preservatives Oily wastes Waste solvents Contaminated containers and packaging Waste batteries Photographic chemicals	Oil/water, hydrocarbon/water emulsions Laboratory chemicals Sludges from treatment operations Tempering salts containing cyanide	Medical wastes Pharmaceuticals waste Halogenated organics PCB containing wastes Waste inks, varnishes Ion exchange residues	Explosive wastes Inorganic wastes that do not contain heavy metals Waste ash and cinder Waste soil and sand Spent catalysts

4.2 Challenges for future research

The literature review carried out and the description of the network structures implemented as multi-objective location and routing problems show how researchers have tried to simulate the true conditions required by the design of a hazardous waste management system. However, addressing these complex systems as mathematical models is a challenging task due to the computational effort required for their solution. Recent studies have not only formulated the traditional location and routing problem to develop hazardous waste management systems but have also incorporated other features such as:

- Modeling waste collection and transportation processes with multiple periods. The foregoing considers that the activities associated with waste management are carried out by periods (daily, weekly, monthly, etc. depending mainly on the needs of the generator and the type of waste).
- Stochastic programming for parameters such as the amount of waste generated at each node. The conditions of uncertainty bring the problem closer to real-life because the amount of waste that can be generated in each source is not known with certainty.

- Multi-objective optimization with objectives other than cost and risk. Most studies of hazardous waste management have been formulated to minimize the total costs and risks to the system. However, environmental or customer satisfaction objectives can also be considered.

According to this, the challenges for future research on this topic require the implementation of the network structures proposed in this study, however, the formulation of mathematical models should focus on applying existing characteristics and conditions in reality. The development of case studies with real applications is a line of research of great importance since most studies are based on models solved from generic data.

5 CONCLUSION

Rapid economic growth and urbanization have caused a waste generation to increase over the years. Within this waste, it is possible to find hazardous wastes, which have characteristics that represent a risk to people who are exposed to them and the environment. The absence of efficient practices for hazardous waste management adversely affects human health and the environment. Therefore, the design of a hazardous waste management system is a fundamental task to guarantee the correct handling of this waste from its collection to its final disposal. The modeling of these systems as multi-objective location and routing problems have been widely studied in the literature since they can solve the location and transport decisions that these systems must define, and in the same way, they allow optimization of objectives such as costs, risks, environmental effects, among others.

Considering that to formulate these models it is necessary to establish an efficient network structure involving all parts of the system, this paper discusses the different frameworks used over the years by researchers to design hazardous waste management systems. The main conclusion of the research is that to achieve an effective system, the following must be defined: the hazardous waste collection processes between the generation nodes; the location of three types of facilities: recycling centers, treatment centers with their type of technology, and disposal facilities and the transportation of waste residue between previously selected facilities. This document proposes a framework that serves as a point of departure for developing a hazardous waste management system, taking into account the characteristics and advances found in the literature so that it can be applied in future research on this topic.

REFERENCES

[1] Samanlioglu, F., A multi-objective mathematical model for the industrial hazardous waste location-routing problem. *European Journal of Operational Research*, **226**(2), pp. 332–340, 2013. DOI: 10.1016/j.ejor.2012.11.019.

[2] Alumur, S. & Kara, B.Y., A new model for the hazardous waste location-routing problem. *Computers and Operations Research*, **34**(5), pp. 1406–1423, 2007. DOI: 10.1016/j.cor.2005.06.012.

[3] Rabbani, M., Heidari, R., Farrokhi-Asl, H. & Rahimi, N., Using metaheuristic algorithms to solve a multi-objective industrial hazardous waste location-routing problem considering incompatible waste types. *Journal of Cleaner Production*, **170**, pp. 227–241, 2018. DOI: 10.1016/j.jclepro.2017.09.029.

[4] Yu, H. & Solvang, W.D., An improved multi-objective programming with augmented ε-constraint method for hazardous waste location-routing problems. *International Journal of Environmental Research and Public Health*, **13**(6), 2016. DOI: 10.3390/ijerph13060548.

[5] Yilmaz, O., Kara, B.Y. & Yetis, U., Hazardous waste management system design under population and environmental impact considerations. *Journal of Environmental Management*, **203**, pp. 720–731. DOI: 10.1016/j.jenvman.2016.06.015.

[6] Asgari, N., Rajabi, M., Jamshidi, M., Khatami, M. & Farahani, R.Z., A memetic algorithm for a multi-objective obnoxious waste location-routing problem: A case study. *Annals of Operations Research,* **250**(2), pp. 279–308, 2017. DOI: 10.1007/s10479-016-2248-7.

[7] Giannikos, I., A multiobjective programming model for locating treatment sites and routing hazardous wastes. *European Journal of Operational Research,* **104**(2), pp. 333–342, 1998. DOI: 10.1016/S0377-2217(97)00188-4.

[8] Zografos, K. & Samara, S., A combined location-routing model for hazardous waste transportation and disposal. *Transportation Research Record,* **1245**, pp. 52–99, 1989.

[9] ReVelle, C. & Jared D.S., Simultaneous siting and routing in the disposal of hazardous wastes. *Transportation Science*, **25**(2), pp. 138–145, 1991.

[10] List, G. & Mirchandani, P., Integrated network/planar multiobjective model for routing and siting for hazardous materials and wastes. *Transportation Science*, **25**(2), pp. 146–156, 1991. DOI: 10.1287/trsc.25.2.146.

[11] Current, J. & Ratick, S., A model to assess risk, equity and efficiency in facility location and transportation of hazardous materials. *Location Science*, **3**(3), pp. 187–201, 1995. DOI: 10.1016/0966-8349(95)00013-5.

[12] Nema, A.K. & Gupta, S.K., Optimization of regional hazardous waste management systems: an improved formulation. *Waste Management*, **19**(7), pp. 441–451, 1999. DOI: 10.1016/S0956-053X(99)00241-X.

[13] Caballero, R., González, M., Guerrero, F.M., Molina, J. & Paralera, C., Solving a multiobjective location routing problem with a metaheuristic based on tabu search: Application to a real case in Andalusia. *European Journal of Operational Research,* **177**(3), pp. 1751–1763, 2007. DOI: 10.1016/j.ejor.2005.10.017.

[14] Zhao, J. & Verter, V., A bi-objective model for the used oil location-routing problem. *Computers and Operations Research*, **62**, pp. 157–168, 2015. DOI: 10.1016/j.cor.2014.10.016.

[15] Zhao, J., Huang, L., Lee, D.H. & Peng, Q., Improved approaches to the network design problem in regional hazardous waste management systems. *Transportation Research Part E: Logistics and Transportation Review*, **88**, pp. 52–75, 2016. DOI: 10.1016/j.tre.2016.02.002.

[16] Zhao, J. & Huang, L., Multi-period network design problem in regional hazardous waste management systems. *International Journal of Environmental Research and Public Health*, **16**(11), 2019. DOI: 10.3390/ijerph16112042.

[17] Rabbani, M., Heidari, R. & Yazdanparast, R., A stochastic multi-period industrial hazardous waste location-routing problem: Integrating NSGA-II and Monte Carlo simulation. *European Journal of Operational Research*, **272**(3), pp. 945–961, 2019. DOI: 10.1016/j.ejor.2018.07.024.

[18] Farrokhi-Asl, H., Makui, A., Jabbarzadeh, A. & Barzinpour, F., Solving a multi-objective sustainable waste collection problem considering a new collection network. *Operational Research*, **20**(4), pp. 1977–2015, 2020. DOI: 10.1007/s12351-018-0415-0.

[19] Delfani, F., Kazemi, A., Seyedhosseini, S.M. & Niaki, S.T.A., A green hazardous waste location-routing problem considering the risks associated with transportation and population. *International Journal of Engineering, Transactions B: Applications,* **33**(11), pp. 2272–2284, 2020. DOI: 10.5829/ije.2020.33.11b.18.

[20] Rabbani, M., Nikoubin, A. & Farrokhi-Asl, H., Using modified metaheuristic algorithms to solve a hazardous waste collection problem considering workload balancing and service time windows. *Soft Computing,* **25**(3), pp. 1885–1912, 2021. DOI: 10.1007/s00500-020-05261-4.
[21] Tirkolaee, E.B., Abbasian, P. & Weber, G.W., Sustainable fuzzy multi-trip location-routing problem for medical waste management during the COVID-19 outbreak. *Science of the Total Environment,* **756**, 2021. DOI: 10.1016/j.scitotenv.2020.143607.
[22] Nagy, G. & Salhi, S., Location-routing: Issues, models and methods. *European Journal of Operational Research*, **177**(2), pp. 649–672, 2007. DOI: 10.1016/j.ejor.2006.04.004.

SECTION 5
ENVIRONMENTAL STRATEGIES AND TACTICS

ATTRIBUTION OF POLLUTION DISCHARGES IN COASTAL WATERS DURING THE COVID-19 LOCKDOWN USING REMOTE SENSING AND BIOINDICATORS

MASSIMILIANO LEGA[1], GABRIELE MEDIO[1], THEODORE ENDRENY[2], GERMANA ESPOSITO[3], VALERIA COSTANTINO[3] & ROBERTA TETA[3]
[1]Department of Engineering, University of Naples Parthenope, Italy
[2]Department of Environmental Resources Engineering, College of Environmental Science and Forestry, State University of New York, USA
[3]Department of Pharmacy, University of Naples Federico II, Italy

ABSTRACT
Attribution of environmental pollution phenomena and their causes represents a real challenge due to the complexity of the scenarios to be examined and the multiple sources to be considered. Furthermore, the definition of the parameters to be monitored, considering the specific phenomenon, becomes very difficult, especially in relation to the available technological solutions and operational needs. Various elements, naturally present in the scenario under examination, can be used as perfect sensors/indicators to detect the status of the environment and effectively reveal the relationships between specific sources of pollution and target areas. This study demonstrates how pollution attribution is achieved using remote sensing of cyanobacteria, which are excellent bioindicators due to their sensitivity to multiple stressors and rapid response to habitat changes throughout the event. The study was conducted in the coastal waters of the Campania region (southern Italy) for the period before, during and after the COVID-19 lockdown. Along the coastal area examined, a special focus was dedicated to Lake Avernus. This program detected the presence of specific cyanobacteria species, some toxic, and revealed the relationship between their presence and anthropic pollution activities. This paper demonstrates how deployment of multidisciplinary analysis with choreographed use of remote/proximal sensing and in situ sampling across space and time utilized the cyanobacteria to characterize the impact of human activities. The study allowed environmental forensic teams to identify and confirm the anthropogenic origin of the environmental variations that stimulate the cyanobacterial blooming.
Keywords: environmental impact assessment, pollution, coastal waters, multispectral analysis, multilayer analysis, proximal sensing, remote sensing, hierarchical monitoring.

1 INTRODUCTION

Identifying a polluted site is an important action for safeguarding health, however often the process of analysing pollution phenomena is limited to verifying the effects and not the causes. Attribution of environmental pollution phenomena and their causes represents a real challenge due to the complexity of the scenarios to be examined and the multiple sources to be considered. Furthermore, the definition of the parameters to be monitored, considering the specific phenomenon, becomes very difficult, especially in relation to the available technological solutions and operational needs.

Understanding a pollution phenomenon often requires a multidisciplinary approach and a multi-temporal analysis, but, unfortunately, sometimes also these types of approaches do not reveal the cause–effect relationships; environmental issues are often faced where there is the evidence of concomitant changes and effects in a specific location and time without immediately understanding the phenomena relationships. An example of these flaws occurred during the COVID-19 lockdown in Italy, when the environment was affected by important changes and the only certainty is that they happened in that particular period and not before and not after. Many people and the media in general have interpreted the environmental changes detected during the lockdown as a consequence of the change of

WIT Transactions on Ecology and the Environment, Vol 257, © 2022 WIT Press
www.witpress.com, ISSN 1743-3541 (on-line)
doi:10.2495/WMEI220151

habits of the communities and therefore hypothesizing them as the causes of the changes. However, much of this news does not arise from real scientific studies and has been disseminated by the media as "sensationalism".

Our previous research studies [1] have already analyzed pollution phenomena along the coasts of the Campania region (Southern Italy) using combined techniques and technologies (remote sensing, proximal sensing and in situ surveys). In particular, a specific approach was developed with a hierarchical monitoring approach called multi-level and multi-parametric monitoring framework (MuM3 hereafter) [2], [3] proved to be particularly effective and efficient in the analysis of source–path–target route of pollution phenomena. When available technologies do not allow to observe some parameters directly, bioindication takes on a fundamental role. Indeed, various elements, naturally present in the scenario under examination, can be used as perfect sensors/indicators to detect the status of the environment and effectively reveal the relationships between specific sources of pollution and target areas; in particular, our research group has investigated the use of cyanobacteria as bioindicators of anthropic pollution [4], [5].

There are several spots of the coastal areas in Campania where cyanobacteria blooms have been detected [6] and in many cases their presence was associated with the eutrophication and, more generically, pollution. Furthermore, we have discovered that cyanobacteria blooms in some lakes along the Campania coasts were linked to anthropogenic pressure and the environmental stress conditions meant that the cyanobacteria also produced toxins dangerous for human health [7].

The aim of this work is to combine and implement the MuM3 and the fast detection strategy (FDS) [8], previously tested and validated, with a top-down approach applied to a real "test bed". Consequently, using both remote and proximal data, beginning at the satellite level and scaling-down at different spatial–temporal resolutions, the validity of the added value of such a method is explored.

The study was conducted in the coastal waters of the Campania region (southern Italy) (Fig. 1) for the period before, during and after the COVID-19 lockdown. In a first step the entire coastal area located in north of Naples was examined, then a special focus was dedicated to Lake Avernus, a volcanic coastal lake, that due to its funnel shape collects rainwater as well as waters of different origins washed out from the surrounding area; as perfect "test bed", Lake Avernus experienced several cyanobacterial blooms in different years, alternating different species and associated toxins; in all cases, the blooming event started in association with stress condition defined from specific environmental status. This study shed light on causes and consequences of the environmental status at the turn of COVID-19 lockdown and deepened the relationship between the variation of anthropic pressure and cyanobacterial bloom events.

2 METHODS

To detect discharges in coastal waters and in general pollution phenomena along the coasts of Campania, in southern Italy, we used a hierarchical monitoring that includes acquisition and analysis of remote data acquired by satellite, proximal data acquired by drone, and in-situ sampling. Consequently, using both remote and proximal data, beginning at the satellite level and scaling-down at different spatial–temporal resolutions, the validity of the added value of such a method was explored. In particular, the combination of two platforms/sensors was used: remote (satellites of the Copernicus Sentinel 2 mission) and proximal (DJI Mavic 2 Pro drone). The twin satellites (2A and 2B) of the Copernicus Sentinel 2 mission are equipped with an optical instrument that samples 13 spectral bands (see Table 1).

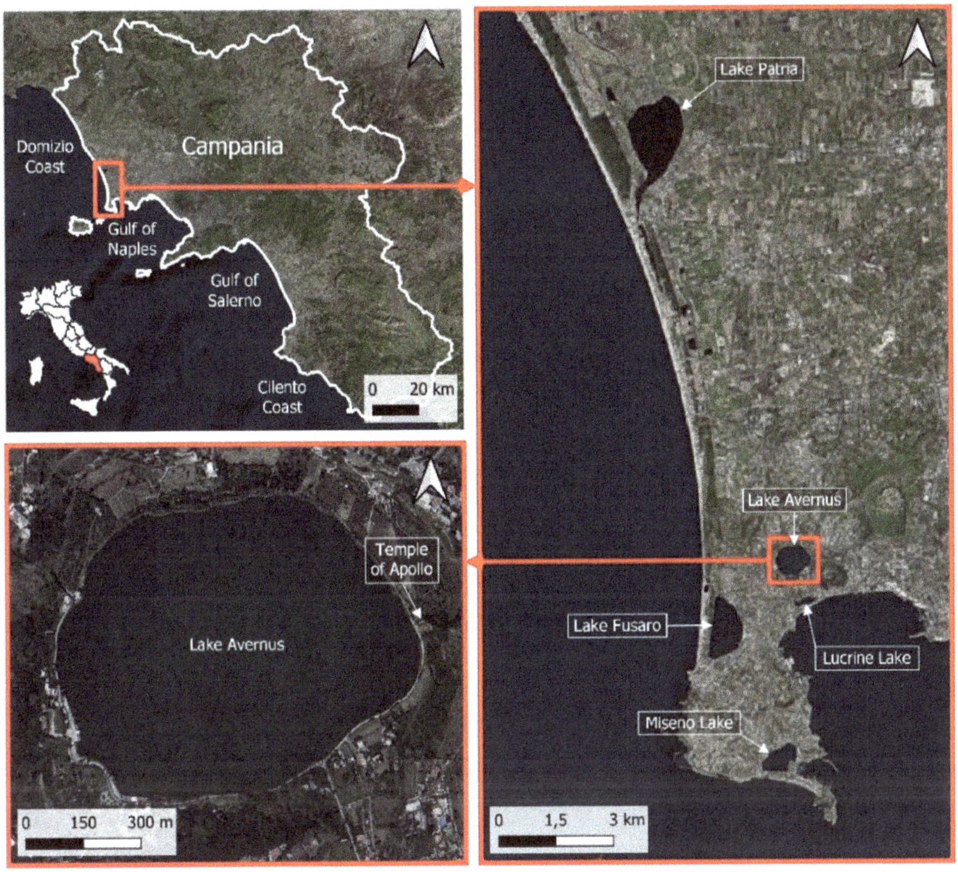

Figure 1: Study sites are to the west of the Italian Gulf of Naples (upper left), along the Domitian Coast (right), and include Lake Avernus (lower left).

The first activity was the satellite-based multispectral analysis of the coastal area over a wide time interval, spanning pre and during the COVID-19 period, from 2019 to 2021.We used the Sentinel 2 data to calculate two multispectral indexes to transform the data into useful information: NDVI (normalized difference vegetation index) and SABI (surface algal bloom index). The calculation of the indices was carried out using the radiometrically corrected Sentinel 2 data, subsequently a classification was performed aimed at highlighting the areas with the plumes of discharges and the inlets of rivers into the sea.

The NDVI is one of the best-known spectral indices and was used for the first time by Rouse et al (1973); it allows to identify the vegetated areas and their vigor by exploiting the reflectance of red (RED) and near infrared (NIR).However, this index is also sensitive to chlorophyll present in algal and bacterial blooms (e.g. cyanobacteria) within coastal waters; therefore, it can be used, with the necessary precautions, also to analyze eutrophic conditions due to the supply of nutrients contained in waste or contaminated water. The NDVI is defined with the following formula:

$$NDVI = (R_{NIR} - R_{RED})/(R_{NIR} + R_{RED}) \quad (1)$$

www.witpress.com, ISSN 1743-3541 (on-line)

Table 1: Spectral bands for the SENTINEL-2 sensors (S2A and S2B) showing spatial resolution (m), central wavelength (nm), bandwidth (nm) and a description, where VNIR is very near infrared and SWIR is shortwave infrared. (*Source: MSI Instrument – Sentinel-2 MSI Technical Guide – Sentinel Online (esa.int).*)

Band number	Spatial resolution (m)	Central wavelength (nm)		Bandwidth (nm)		Description
		S2A	S2B	S2A	S2B	
1	60	442.7	442.3	21	21	Ultra blue
2	10	492.4	492.1	66	66	Blue
3	10	559.8	559.0	36	36	Green
4	10	664.6	665.0	31	31	Red
5	20	704.1	703.8	15	16	VNIR
6	20	740.5	739.1	15	15	VNIR
7	20	782.8	779.7	20	20	VNIR
8	10	832.8	833.0	106	106	VNIR
8a	20	864.7	864.0	21	22	VNIR
9	60	945.1	943.2	20	21	SWIR
10	60	1373.5	1376.9	31	30	SWIR
11	20	1613.7	1610.4	91	94	SWIR
12	20	2202.4	2185.7	175	185	SWIR

The SABI allows the detection of biomass floating on water and is therefore particularly sensitive to algal and bacterial blooms. The SABI combines the reflectance of the red (RED), green (GREEN), blue (BLUE) and near infrared (NIR) spectral bands:

$$SABI = (R_{NIR} - R_{RED})/(R_{BLUE} + R_{GREEN}) \quad (2)$$

Thanks to the data acquisitions of the sensors transported aboard platforms located at different altitudes, the study was conducted on the basis of different relationships between the spatial resolution of the image and the size of the detected area.

The Copernicus Sentinel 2 mission provides the user segment with a collection of orthophotos, 100×100 km^2 in UTM / WGS84 projection, relating to the 13 spectral bands listed above. In particular, bands 2, 3, 4 and 8 were used to calculate the NDVI and SABI indices, consequently the spatial resolution of the satellite data is 10 m.

The drone DJI Mavic 2 Pro, on the other hand, is UAV (unmanned aerial vehicle) equipped with a fully stabilized three-axis gimbal RGB camera and capable of taking 20 megapixel photos. In addition, we have added to the standard payload a multispectral camera (MAPIR Survey 3) to permit acquisition of other three bands ("Orange", "Ciano", "Near Infrared"). Using the RGB camera bands it was possible to have a proximity confirmation, in the visible range, of what was observed remotely. In our study the proximal acquisitions have a high spatial resolution that range from a few cm/px, depending from the altitude of flight path (e.g., we acquired a large part of images at spatial resolutions of 5–10 cm).

The data acquired were processed using the QGIS software; with this program the various spectral indices were calculated and, through the classification of the index values, thematic maps were created in order to visually enhance certain aspects of the phenomenon of interest. Particular attention was paid to maintaining constant the subdivision into classes for all the elaborations carried out to make the results and, above all, the relative representations in the form of thematic maps, comparable.

3 RESULTS AND DISCUSSION

The multispectral analyses on the coasts of Campania, in particular the Gulf of Naples and the Domitian coast, were carried out using satellite data over the time interval 2019–2021, divided into years and months (Fig. 2). The definition of classes and colours represented the crucial step in terms of efforts and time, as in the case of all such analyses, in that only a correct classification allows to bring out the areas of interest.

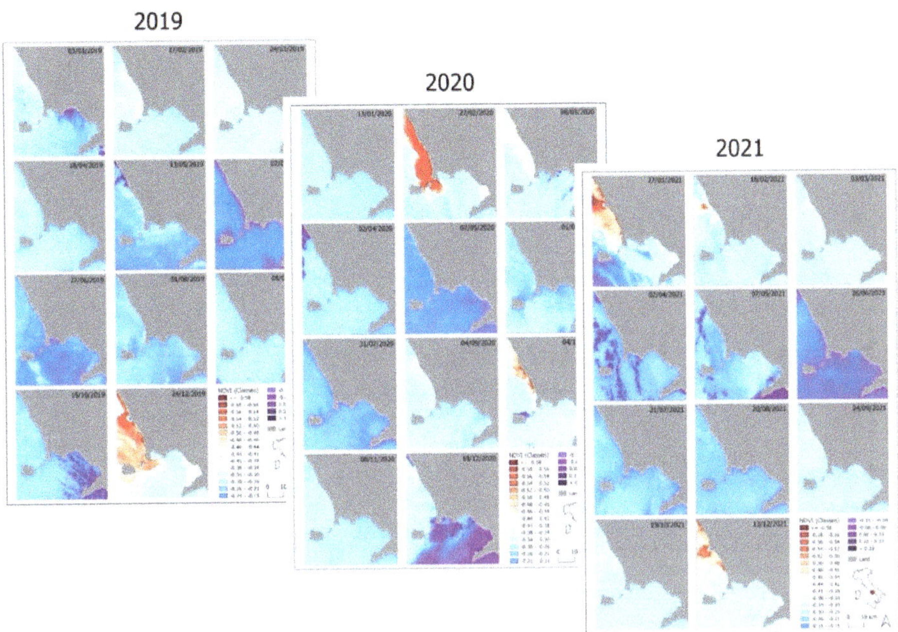

Figure 2: Thematic maps (grey is land area, color is water area) obtained from the calculation of the NDVI (normalized difference vegetation index) and subsequent classification of the results in 20 classes for years 2019 to 2021. The purple and red colors identify the waters of greatest interest; in particular, those in purple identify a potential presence of chlorophyll a (not to be confused with spectral response of atmospheric disturbances, i.e. clouds), those in red adjacent to the coast are near the mouths of rivers and canals and are related to the plumes deriving from periods of intense rain.

In the NDVI thematic maps of Campania coasts, the purple and red zones identify specific characteristics of the areas. Those map areas in purple highlight the greater concentration of chlorophyll *a* and, therefore, delimit water potentially impacted by environmental issues due to the input of nutrients (for example sewage drains, collection channels of irrigation water from agricultural fields, etc.), which give rise to eutrophication phenomena. Those map areas in red highlight the water impacted by the introduction of waters deriving from intense periods of rain and often with even greater turbidity. Since the observed area were very large while the extension of the above issues was limited in size and sometimes hidden by atmospheric phenomena, we decided to focus on a specific target area that had best showed the relationships between anthropogenic pressure conditions and the environmental response, the Lake Avernus (Fig. 3).

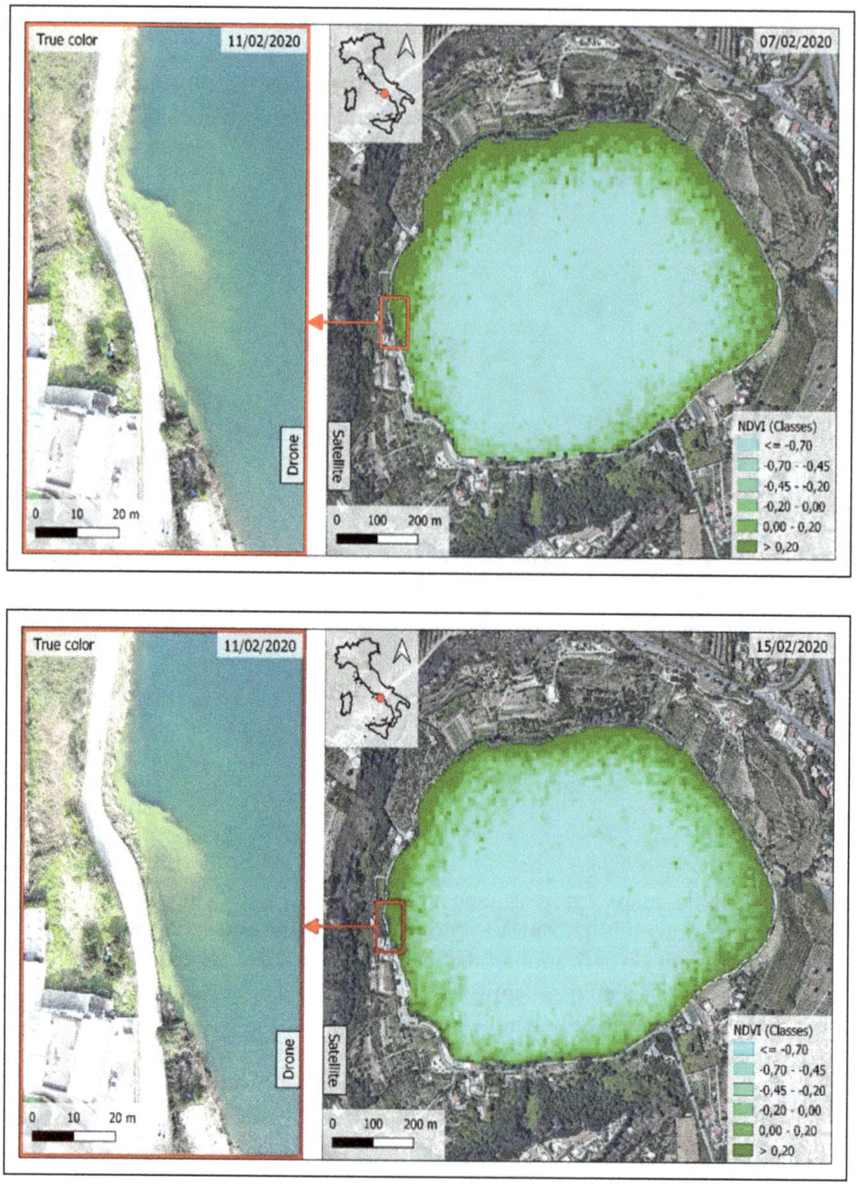

Figure 3: Drone (left) and satellite (right) thematic maps of the Lake Avernus study site for NDVI (normalized difference vegetation index) triggered attribution. The multispectral satellite images were obtained 7 February 2020 (top right) and 8 days later (bottom right) and used to calculate the NDVI and subsequent classification of the results in VI classes. A higher NDVI (dark green) corresponds to higher chlorophyll a content. The drone image in true color has the spatial resolution to reveal the plume discharge, and was obtained on 11 February 2020, after the NDVI analysis of the 7 February 2020 image revealed pollution concern; the drone can attribute the green plume to the cyanobacterial bloom.

The analysis was carried out on the lake using the NDVI and SABI indices in February 2020, when the lake was interested from a *Microcystis aeruginosa* bloom event [7]. The multispectral data refer to the dates closest to the sampling campaign (see below for details). During the sampling, a drone mission was also performed to survey proximal data.

As evident, there is a significant correspondence of blooming peak along the shores of the lake between satellite thematic map and drone RGB images. Driven by the positive results, the satellite data analysis was extended to the years 2018, 2019, 2020 and 2021 and for each year to at least once a month, choosing from the available dates those with the lowest cloud cover. A synoptic table that brings together all the thematic maps is shown in Fig. 4.

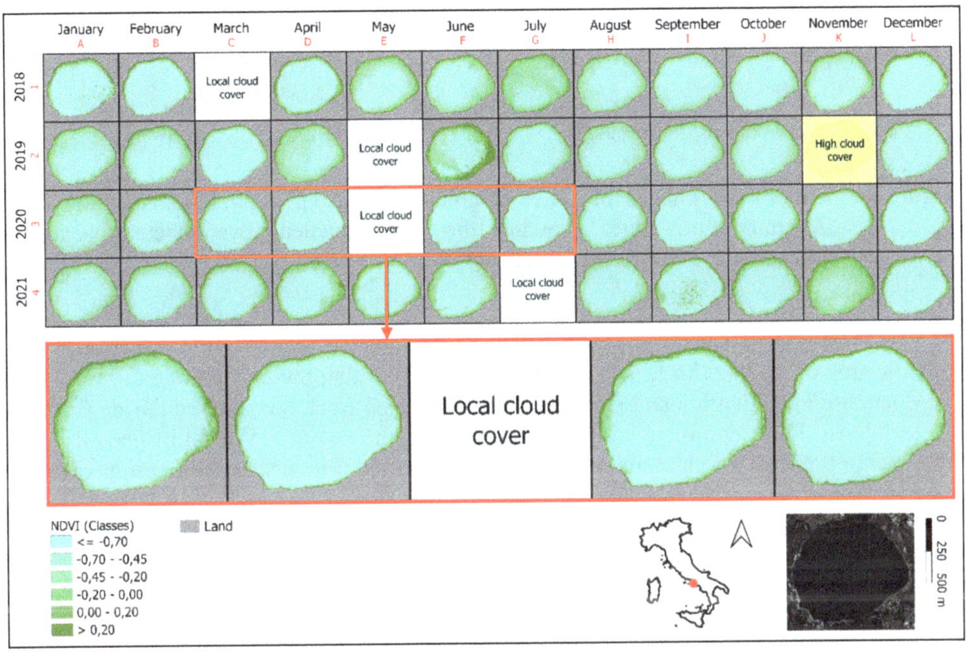

Figure 4: Time series as a synoptic table of the NDVI (normalized difference vegetation index) thematic maps of the Lake Avernus study site for all months in the years 2018–2021, where grey is land, magenta is water with no detected algal bloom concern, and dark green is maximum algal bloom concern.

The periodicity of blooming is certainly evident and reflects aspects of seasonality both connected to the climate and to the human activities of the areas around the lake.

The study of cyanobacterial species throughout the parallel analysis of samples taken in situ revealed the presence of two prevalent species: *Microcystis aeruginosa* and *Planktothrix rubescens* (Fig. 5). Studies on the reasons why one species or the other one blooms are still ongoing. Both species are potentially toxic. The link between toxin production and environmental stress has been previously ascertained [7].

In the period March to July 2020, the magnitude of the blooming was not comparable to the homologous periods of the other annuities; the extent of the perimeter bloom was remarkably reduced in that period corresponding to the first and most severe Italian COVID-19 lockdown.

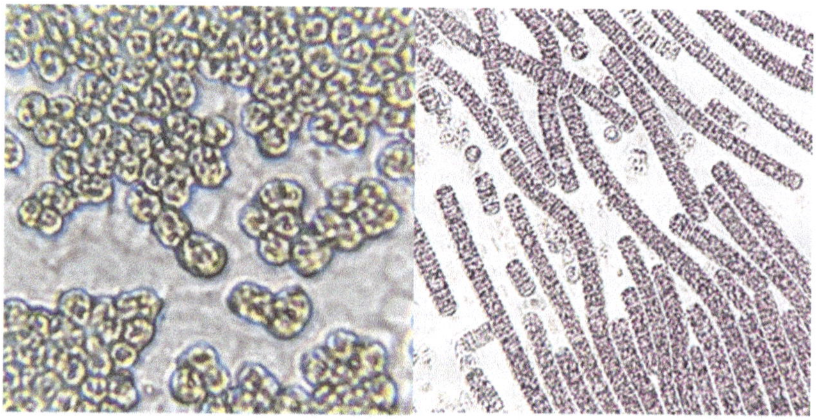

Figure 5: Laboratory images of water samples taken from algal blooms in the Lake Avernus study site, showing the two prevalent cyanobacterial species, *Microcystis aeruginosa* and *Planktothrix rubescens*.

This condition is justified by the reduced anthropogenic pressure in that period on the specific area of the lake. In fact, it should be pointed out that the area around the lake has a prevalent use in cultivation and recreation, both of which were very limited during COVID-19 lockdown. For a complete and in-depth analysis, the use of SABI multispectral index, specific for the study of algal bloom was also investigated. Fig. 6 shows the results obtained from the application of SABI index to the satellite images of Lake Avernus in the same dates used for NDVI index for comparison. As clearly visible no differences are observed between the two sets of multispectral data, confirming the previous results and also the presence of cyanobacteria.

4 CONCLUSIONS

This paper demonstrates how the deployment of multidisciplinary analysis with composed use of remote/proximal sensing and in situ sampling across space and time, makes possible to utilize cyanobacteria to characterize the impact of human activities.

This study confirms definitely how pollution attribution can be achieved using remote sensing of cyanobacteria, which are excellent bioindicators due to their sensitivity to multiple stressors and rapid response to habitat changes throughout the event.

This study allowed also to identify and confirm the anthropogenic origin of the environmental variations occurred during COVID-19 lockdown in Italy, that stimulate the cyanobacterial blooming.

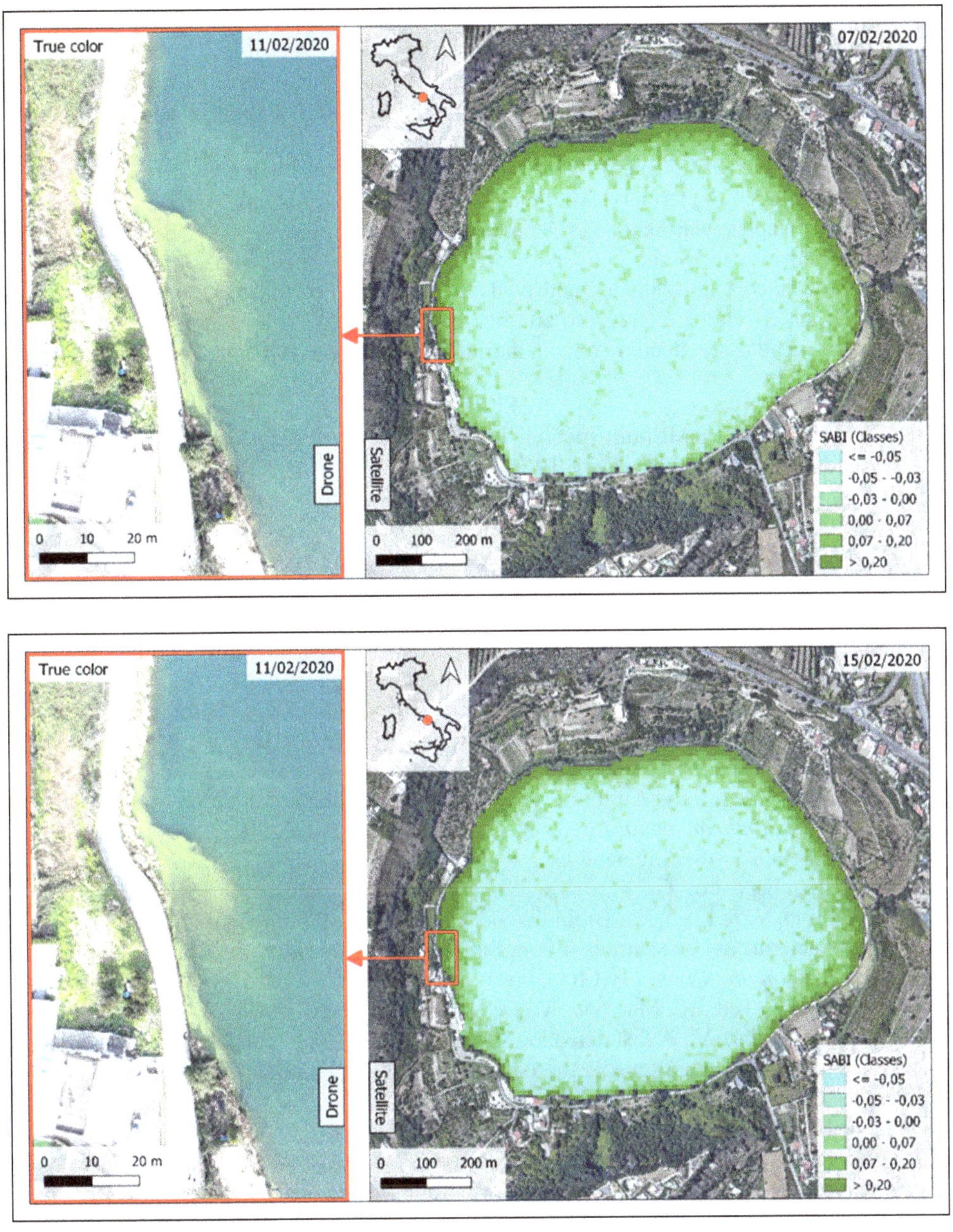

Figure 6: Drone (left) and satellite (right) thematic maps of the Lake Avernus study site for SABI (surface algal bloom index) triggered attribution. The SABI analysis and subsequent classification resulted in VI classes, and no differences with NDVI (normalized difference vegetation index) were noted (see Fig. 3). The satellite image of 7 February 2020 (top right) allowed calculation of the SABI, which identified a potential algal bloom, and triggered the drone flight and true color picture (left images) showing the green discharge associated with the cyanobacterial bloom.

ACKNOWLEDGEMENTS

The data presented in this work are derived from the project "Attivita pilota di Monitoraggio di Cianobatteri nella fascia costiera della regione Campania" funded by the "Centro di Riferimento Regionale per la Sicurezza Sanitaria del Pescato (CRiSSaP)" and performed in cooperation with the Agency for the Environmental Protection of Campania region in Italy (ARPAC), University of Naples "Federico II" (Department of Pharmacy, ref. V. Costantino and Department of Veterinary Medicine and Animal Production, ref. A. Anastasio), University of Naples "Parthenope" (Department of Engineering, ref. M. Lega). This project was supported in part by the University of Naples "Parthenope" (Italy) under "Bando di sostegno alla ricerca individuale" and by Department of Pharmacy, University of Naples Federico II under "Bando contributo alla ricercar—Anno 2021" for the project "CyaAq—I cianobatteri come promettente risorsa di antimicrobici non convenzionali".

REFERENCES

[1] Errico, A. et al., SAR/multispectral Image fusion for the detection of environmental hazards with a GIS, 2014. DOI: 10.1117/12.2066476.

[2] Lega, M., Casazza, M., Teta, R. & Zappa, C.J., Environmental impact assessment: A multilevel, multi-parametric framework for coastal waters. *International Journal of Sustainable Development and Planning*, **13**(8), pp. 1041–1049, 2018. DOI: 10.2495/SDP-V13-N8-1041-1049.

[3] Lega, M., Kosmatka, J., Casazza, M., De Maio, L., Severino, D. & Teta, R., Environmental monitoring and assessment: A multi-scale and multiparameter case study in Campania region (S. Italy) on the recreational seawater quality evaluation. *Regional Studies in Marine Science*, **39**, 2020. DOI: 10.1016/j.rsma.2020.101465.

[4] Teta, R., Romano, V., Della Sala, G., Picchio, S., De Sterlich, C., Mangoni, A., Di Tullio, G., Costantino, V. & Lega, M., Cyanobacteria as indicators of water quality in Campania coasts, Italy: A monitoring strategy combining remote/proximal sensing and in situ data. *Environmental Research Letters*, **12**(2), 024001, 2017. DOI: 10.1088/1748-9326/aa5649.

[5] Teta, R., Esposito, G., Casazza, M., Zappa, C.J., Endreny, T.A., Mangoni, A., Costantino, V. & Lega, M., Bioindicators as a tool in environmental impact assessment: cyanobacteria as a sentinel of pollution. *International Journal of Sustainable Development and Planning*, **14**(1), pp. 1–8, 2019. DOI: 10.2495/SDP-V14-N1-1-8.

[6] Esposito, G., Teta, R., Marrone, R., De Sterlich, C., Casazza, M., Anastasio, A., Lega, M. & Costantino, V., A fast detection strategy for cyanobacterial blooms and associated cyanotoxins (FDSCC) reveals the occurrence of Lyngbyatoxin A in Campania (south Italy). *Chemosphere*, **225**, pp. 342–351, 2019. DOI: 10.1016/j.chemosphere.2019.02.201.

[7] Teta, R., Esposito, G., De Sterlich, C., Lega, M. & Costantino, V., Early detection of cyanobacterial blooms and associated cyanotoxins using fast detection strategy. *Journal of Visualized Experiments*, **2021**(168), 2021. doi: 10.3791/61889.

[8] Teta, R., Sala, G.D., Esposito, G., Stornaiuolo, M., Scarpato, S., Casazza, M., Anastasio, A., Lega, M. & Costantino, V., Monitoring cyanobacterial blooms during the COVID-19 pandemic in Campania, Italy: The case of Lake Avernus. *Toxins*, **13**(7), 2021. DOI: 10.3390/toxins13070471.

CONSTRAINT MAPPING FOR AVOIDING ADVERSE EFFECTS OF DEVELOPMENT: THE APPLICATION TO POINT WILSON, AUSTRALIA, AND ITS AFTERMATH

BRYAN R. JENKINS*
University of Adelaide, Australia

ABSTRACT

Avoidance of impacts on areas of environmental significance is a desired outcome of environmental planning. This paper describes the process of constraint mapping that was used in the siting and layout for a proposed petrochemical plant at Point Wilson, near Geelong, in the Australian State of Victoria. Saltmarsh vegetation on part of the proposed site was part of the prime wintering habitat for the orange-bellied parrot, a critically endangered species. Comprehensive field investigations were undertaken to determine the habitat areas and flight paths of the parrot. Other environmental factors were also defined and spatially mapped. This constraint mapping led to revising the siting and layout of the proposed petrochemical plant. As part of the rezoning process, a buffer zone was established between the industrial development zone and the conservation reserve to protect habitat areas and flight paths. The aftermath of the rezoning is considered. The petrochemical plant did not proceed on economic grounds and subsequent proposals by others for industrial development have not proceeded. The site is currently a basalt quarry. Parrot habitat protection has influenced all of these decisions. The conservation reserve became part of a Ramsar-listed site for flora and fauna protection. However, parrot numbers have declined. While there has been some degradation of the saltmarsh habitat, the decline is attributed to problems of the small population of orange-bellied parrots. A breeding programme is assisting in the recovery of parrot numbers and the conservation reserve is a key component of the reintroduction of parrots from the breeding programme into the wild.

Keywords: constraint mapping; environmental impact assessment; endangered species management.

1 INTRODUCTION

This paper considers the contribution of constraint mapping to environmental impact assessment for the avoidance of adverse impacts of proposed developments. After the introduction, the case study of the application of constraint mapping for the siting and plant layout of a proposed petrochemical plant at Point Wilson, Victoria, Australia, is presented. First the concept of constraint mapping is described followed by a brief summary of the proposed development and the surrounding land uses. While the site is well buffered from residential development, it partly covers the prime wintering habitat of the orange-bellied parrot, a critically endangered species. Next, the studies of the orange-bellied parrot are summarised.

The information on the parrot, other environmental issues, and land use compatibility considerations are discussed in relation to the constraint they pose to each major element of the plant. A siting issues matrix is set out linking constraints to the plant elements. The final step in the constraint mapping is the preparation of composite overlays. For each factor the degree of constraint is shown as grey shading for each plant element. Then, for each plant element, the constraints are combined with dark areas showing the greatest constraint and light areas showing preferred locations to avoid or minimise environmental impacts. The final outcome of the constraint mapping was a relocating of the proposed development to avoid parrot habitat and providing a buffer zone between the plant and habitat areas. This

* *ORCID: http://orcid.org/ 0000-0002-4031-101X*

WIT Transactions on Ecology and the Environment, Vol 257, © 2022 WIT Press
www.witpress.com, ISSN 1743-3541 (on-line)
doi:10.2495/WMEI220161

was reflected in the rezoning decision to allow industrial zoning over part of the site and defining a rural conservation zone as a buffer to habitat areas.

The paper then considers the aftermath of the rezoning decision both in terms of industrial development and orange-bellied parrot management. While the petrochemical plant did not proceed, Point Wilson was considered for a bulk chemical storage and an expansion of the armaments complex and finally a basalt quarry. A recovery programme was established for the management of the orange-bellied parrot as an endangered species.

The paper concludes by highlighting the value of environmental impact assessment in managing the compatibility between development and habitat protection. The concluding comments also note the limitations of environmental impact assessment with respect to species recovery. There is a need for a programme of proactive interventions across all facets of parrot ecology if species survival is to be achieved.

2 THE APPLICATION OF CONSTRAINT MAPPING TO POINT WILSON

2.1 The concept of constraint mapping

One of the most valuable tools that can be used in environmental planning is constraint mapping. This concept was developed by Ian McHarg as set out in his book *Design with Nature* [1]. It consists of identifying both natural and social processes that need to be considered in designing proposed developments, in addition to the technical and economic considerations in project design.

At the plan development stage, the approach involves identifying the critical factors and ranking them from high to low in terms of their ecological or social value. For example, areas of high ecological significance such as endangered species habitat would receive a high ranking. The factors for each value are mapped with areas of higher value shown with darker tone. Where there are multiple factors, the maps are superimposed. The areas which are of high ecological and social value are shown as dark tones while areas with lower social and ecological value are revealed by the lightest tone. This enables the siting of facilities to be undertaken to avoid areas of high ecological and social value.

2.2 Proposed development at Point Wilson and surrounding land use

The process of constraint mapping was used in the planning stages of a proposed petrochemical plant at Point Wilson near Geelong. In the late 1970s, ICI Australia Ltd sought rezoning of nearly 800 ha of land at Point Wilson. The intention was to develop the site for a petrochemical chlor–alkali manufacturing complex. The ICI plant at Botany in New South Wales was being encroached by urban development and ICI was seeking a site that would be buffered from urban development. The land at Point Wilson was surrounded by Avalon Airfield, the Melbourne Metropolitan Board of Works sewage farm, Cheetham Salt Works, and a Commonwealth explosives reserve (Fig. 1). Although the site was remote from residential development, the remnant coastal saltmarsh vegetation was the prime wintering habitat for a critically endangered species – the orange-bellied parrot.

2.3 Studies of the orange-bellied parrot

Issues of concern in relation to the impact of the proposed development on the orange-bellied parrot were:

- Land development affecting habitat areas of significant bird species;

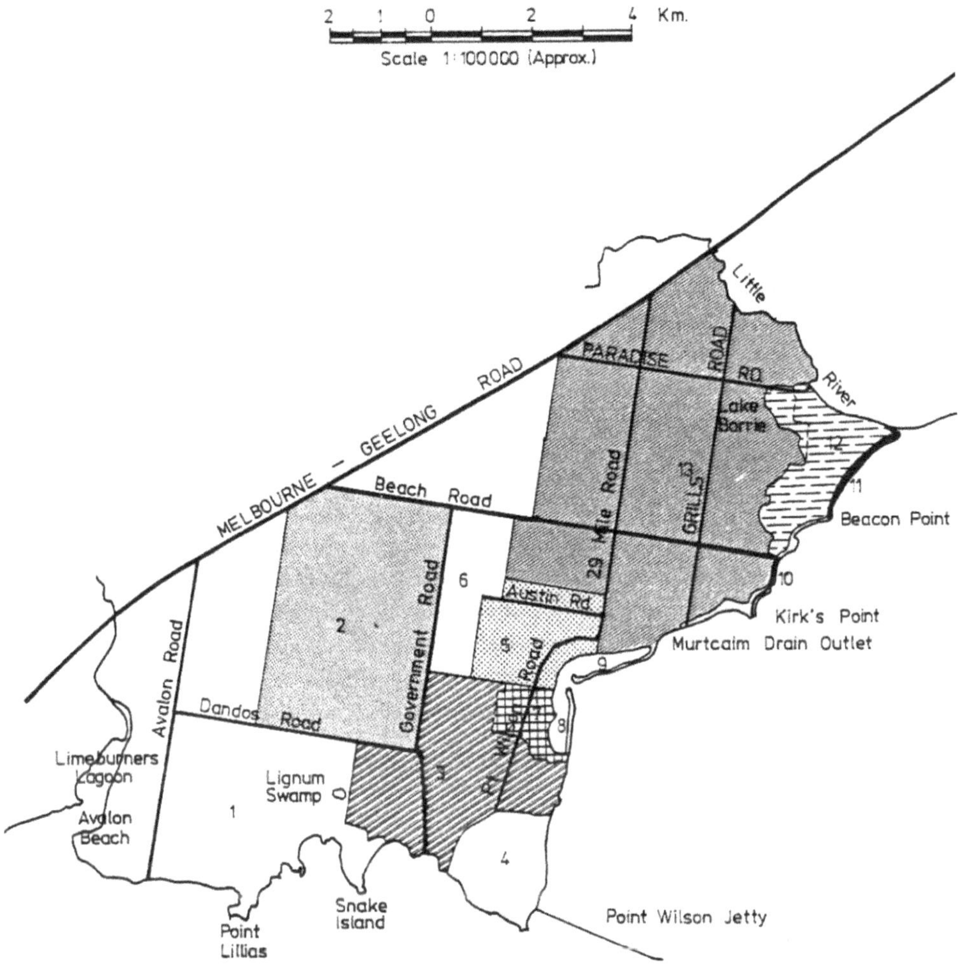

1. CHEETHAM (AVALON) SALTWORKS
2. AVALON AIRFIELD
3. I.C.I. AUSTRALIA POINT WILSON SITE
4. POINT WILSON EXPLOSIVES AREA
5. AUSTIN ROAD LAGOON AREA
6. COMMONWEALTH AREA
7. POINT WILSON SALTMARSH AND FLOODWATER
8. SOUTH SPIT
9. NORTH SPIT
10. KIRK POINT COASTAL STRIP
11. BEACON POINT TO LITTLE RIVER COASTAL STRIP
12. LAKE BORRIE AND ADJACENT LAGOONS
13. M.M.B.W. FARM PASTURE AND FILTRATION PADDOCKS

N

Figure 1: Point Wilson study area [2].

- High structures affecting zones of movement of significant bird species;
- Noise from plant operations disturbing bird usage of the area; and
- Plant emissions that are phytotoxic to plants used as food by significant bird species.

As part of the environmental impact assessment studies for the proposed plant [3] a detailed assessment was undertaken of the potential impacts and the significance of the nearby habitat of the orange-bellied parrot. This began with a review of previous sightings, preferred habitat, known range and ecology (e.g., [4]). A series of field programmes were undertaken to provide a more complete picture of the parrot's ecology, habitat, and range. Continuous field observations of the parrot were undertaken around The Spit near Point Wilson. Initially, this work focussed on feeding and roosting areas (Fig. 2), parrot movements (Fig. 3) and general behaviour patterns (e.g., [2]). This work was extended to assess reaction to disturbance and effects of grazing on bird utilisation of habitat.

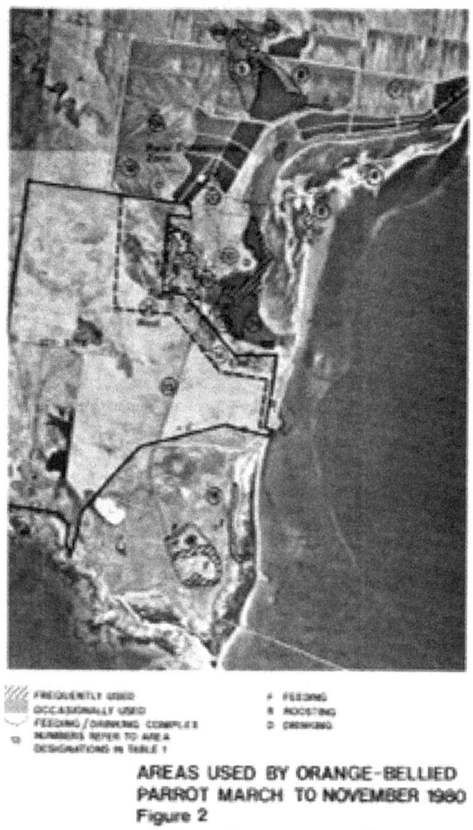

Figure 2: Areas used by orange-bellied parrot.

Figure 3: Zones of movement of orange-bellied parrot [2].

The significance of The Spit as orange-bellied parrot habitat was the presence of saltmarsh. A survey of all coastal saltmarshes in Victoria was undertaken [5]. The floristic and structural diversity of the salt marshes appeared to be important as a determinant of significance of orange-bellied parrot habitat. This was consistent with the field observations at Point Wilson which had shown that the parrot used a wide range of food sources because of their varying seeding times.

Also, two simultaneous total population counts throughout Southeast Australia were undertaken by over 100 observers across the parrot's wintering habitat. This indicated a total population of orange-bellied parrots of less than 100 with the Point Wilson area being the prime location both in numbers and presence throughout the wintering period [6]. Investigations were also undertaken of the summer breeding area in Southwest Tasmania around the Melaleuca Lagoon area of Port Davey [7].

The hydrology of the area was investigated to determine if drainage requirements of the development might adversely affect the ecology of the habitat [8]. The change in drainage inflows due to the proposed development were found to have an insignificant effect on the system hydrology. However, it was found that an existing road embankment on the Point Wilson access road had reduced saltwater inflow to the upper saltmarsh and that a groundwater mound from nearby sewage lagoons was aiding the invasion of freshwater plant species into the saltmarsh community.

Various conservation measures were also studied. Field observations showed that sheep and rabbit grazing had a detrimental impact on habitat. The major areas were fenced to prevent sheep grazing. The studies of disturbance indicated that "sharp" rather than loud noises and visible intrusions were the most significant causes of disturbance. In the proposed development a buffer distance was provided, screen planting begun, and the development designed so that passive storage areas were on the boundary nearest the habitat areas.

2.4 Siting issues matrix

A siting issues matrix (Table 1) was developed between each major element of the petrochemical plant and the key environmental and land use compatibility factors that had been identified [9]. In addition to the concerns about the orange-bellied parrot, other factors that were considered related to adjacent land uses: the maintenance of clearance surfaces for the air traffic movement at Avalon Airfield, the effect of air emissions on sewage farm operations, the waste absorbing capacity of the receiving waters of Port Phillip Bay, and safety distances from explosives storage. The factors considered were:

- areas of orange-bellied parrot habitat – to be avoided for all plant elements;
- zones of movement of the orange-bellied parrot – to be avoided for plant elements greater than 30 m in height;
- habitat areas of other avifauna – to be avoided where possible for all plant elements;
- flight paths of other avifauna – to be avoided where possible for plant elements greater than 30 m in height;
- catchment areas of the saltmarsh – to be kept clear of plant elements from which associated spills of liquid chemicals could occur;
- areas of high permeability – to be kept free of sealed plant areas in order to maintain infiltration inflow to near-surface aquifers;
- landscaped areas set aside for regeneration or plantings – to be kept free of plant elements;
- Avalon Airfield clearance surfaces – tall structures (over 53 m) to be located beneath clearance surface for air traffic;

- Point Wilson explosives area safety distances – no plant elements to be located within the safety distances of the magazines;
- areas sensitive to air emissions – location of plant elements that emit toxic or phytotoxic emissions to be located chosen to minimise air emission concentrations in saltmarsh and sewage farm;
- areas sensitive to noise – plant elements with higher noise levels (especially high frequency noise) to be located as far as possible from parrot habitat.

Table 1: Environmental factors relevant to location of petrochemical plant elements [9].

Environmental factors	Polythene	Polypropylene	Administration	Hydrocarbon storage	Liquid chemical storage	Caustic chlorine	Workshop	Ethylene	Derivative products	Chlorinated hydrocarbons	Boiler house	Salt storage	Liquid waste treatment
Parrot habitat	●	●	●	●	●	●	●	●	●	●	●	●	●
Parrot movement zones	●	●						●			●		
Other avifauna habitat	●	●	●	●	●	●	●	●	●	●	●	●	●
Flight paths	●	●						●					
Saltmarsh catchment				●	●			●		●			
Permeable areas	○	○	○			○	○		○		○	○	○
Landscaped areas	●	●	●	●	●	●	●	●	●	●	●	●	●
Avalon clearance surfaces		●											
Explosives safety distances	●	●	●	●	●	●	●	●	●	●	●	●	●
Air emission sensitivity	○			○		●		●		●	●		
Noise sensitivity						○		●	●	○	○		

● Factor of some significance; ○ Factor of less significance.

2.5 Composite overlays

For each environmental factor a map was prepared indicating sensitive areas by shadings in grey tones. Areas of greatest sensitivity were shaded in the darkest tone: for example, for the permeability map, areas of 10^{-1}m/d permeability were shaded darker than areas of 10^{-2}m/d, while areas of 10^{-3}m/d were left clear. By overlaying the maps of environmental factors that are relevant to the siting of a particular plant element, it is possible to identify the location that has minimum environmental impact: this is the location with the lightest tone. The darker areas indicate locations of relatively greater impact.

A composite set of overlays was prepared for each plant element. In making plant layout decisions, functional, economic, technical and safety factors were also relevant. This can place limits on selecting plant elements on environmental grounds. Comparing the relative tones with the proposed locations gives an impression of the extent to which environmental impacts have been minimised through plant layout.

Certain plant elements warranted greater consideration than others, such as the caustic chlorine plant. In order to use the site to best advantage in environmental terms, such plants are located before less critical elements were located. The caustic chlorine plant (Element 5 in Fig. 4) was located in a relatively pale grey area from an environmental impact viewpoint. Note that Element 12 is the waste treatment facility, which had already been located to provide the most advantageous outfall location.

There were different environmental constraints for other plant elements. For example, the administration complex (Element 2) was less constrained. From the areas of low impact, a location at the main entrance to the site was selected on functional grounds (Fig. 5).

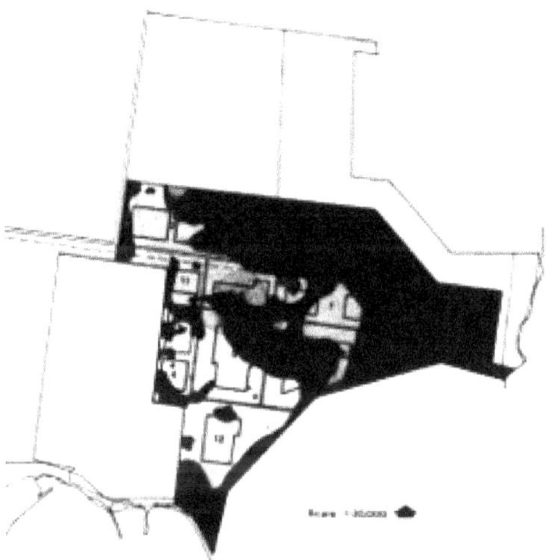

Figure 4: Composite overlay for caustic chlorine plant.

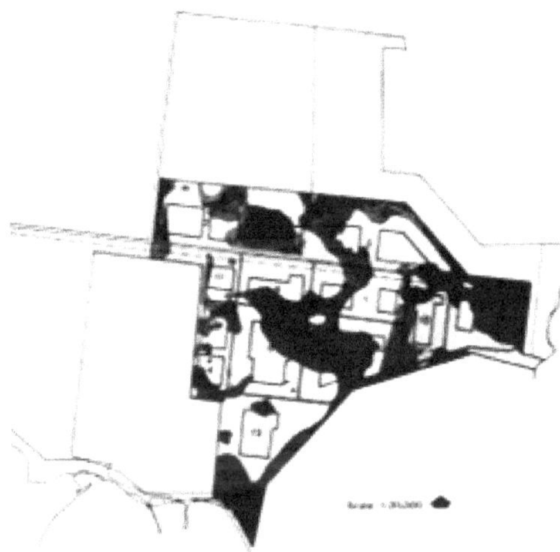

Figure 5: Composite overlay for administration complex [9].

By incorporating environmental and land use compatibility factors in the site layout decision making, it was possible to reduce environmental effects without compromising functional, economic, and other siting considerations. Furthermore, in the land rezoning to industrial the provision was made for a buffer zone between the industrial land and the public conservation zone – shown as "Rural Conservation Zone" in Fig. 6.

3 AFTERMATH

3.1 Industrial development

The rezoning of the ICI site at Point Wilson to industrial use was approved together with a buffer zone around the margins of the Spit Conservation Reserve zoned Rural Conservation. However, the ICI proposal for a petrochemical plant did not proceed on economic grounds.

In the 1990s, the site was considered by the Victorian Government as a bulk liquid storage terminal for hazardous chemicals. In 1991, a major fire at the Coode Island facility for

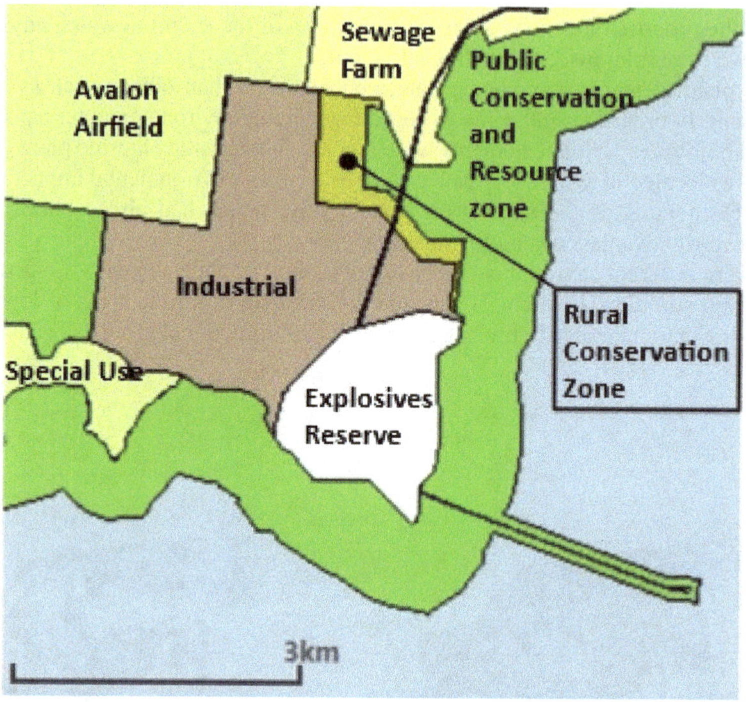

Figure 6: Land use zoning at Point Wilson (Greater Geelong Planning Scheme).

chemical storage in Melbourne led to a review of the Coode Island facility and a recommendation for siting the facility at West Point Wilson (Point Lillias). The Environment Effects Statement for the proposal concluded that construction of the facility would disturb the orange-bellied parrot and other bird species [10]. Following the Ministerial determination on the Environmental Effects Statement, the Victorian Government announced that the proposed bulk chemical storage at Point Wilson would not proceed but the Coode Island operation would be reviewed to ensure it fully meets the required environmental and safety standards.

In 1994, the Commonwealth Government nominated Point Wilson as the site for the East Coast Armaments Complex. There has been an explosives storage and jetty there since 1962. The inquiry into the proposal acknowledged the significance of the nearby parrot habitat [11]. The recommendations of the Inquiry resulted in the erection of a predator-proof fence around the eastern and southern saltmarsh habitat and the installation of more culverts beneath the Point Wilson access road to improve tidal penetration to the upper saltmarsh. The proposal has been deferred.

The former ICI site is currently (2022) operating as a basalt quarry. An expansion of quarry operations has been subject to environmental assessment [12]. The proposed expansion avoids parrot habitat, movement areas and the buffer zone. The indirect environmental issues relating to possible impacts on parrot habitat concerned effects on the catchment of The Spit including freshwater diversion and lowering of groundwater from dewatering to extract basalt deposits below the water table. These issues were considered minor compared to seawater inundation to maintain the upper saltmarsh vegetation. A monitoring and mitigation strategy in relation to these effects was recommended [13].

3.2 Orange-bellied parrot management

The initial work for ICI in 1979 prompted the conservation agencies of Victoria, Tasmania, South Australia, and the Commonwealth together with representatives from ICI, World Wide Fund for Nature (WWF), Birds Australia and Birdlife International to agree to undertake research into the status and ecology of the orange-bellied parrot [14].

The Spit Nature Conservation Reserve and substantial areas of the western shoreline of Port Phillip Bay and the Lake Connewarre wetland complex on Bellarine Peninsula were designated a Ramsar site in 1982. This was not only because of the area's importance for orange-bellied parrots but also for wetlands and waterbirds [15].

In 1983, the Orange-Bellied Parrot Recovery Team was established to coordinate conservation activities and to guide implementation of a recovery plan. The first recovery plan was prepared in 1984. This was the first significant management plan for an endangered species in Australia. The first trial captive breeding population was established in 1986 [14]. Updated plans have been developed every five years with the most recent produced in 2016 [16].

More than one hundred years ago there were concerns for the plight of the species [17] and nearly 60 years ago the rarity of the parrot was documented [18]. However, no actions had been taken to address the decline until the recovery plans were developed and implemented.

Coordinated winter surveys were continued, relying on the contributions of more than a hundred volunteer bird observers. The number of individual parrots counted has declined from 70–90 individuals in the 1980s to fewer than 20 each year since 2001 and fewer than 10 in the 2011–2015 period. In 1992, the parrot was classified as "endangered" and reclassified as "critically endangered" in 2006 under the Environmental Protection and Biodiversity Conservation Act.

Potential wintering habitat loss was investigated as a possible cause of decline in numbers. However, little change in habitat was found since 1983 with most of the gross habitat loss occurring before then [19]. The first population viability analysis showed that high juvenile mortality was the greatest limitation on population increase. The release of 264 captive-bred birds from 1994 to 2006 at the parrot's breeding site failed to re-establish a viable wild population [14].

In 2010, there were indications that extinction of the parrot in the wild was likely in 3–5 years. Immediate action was taken to bolster the captive population into an effective insurance population as well as a source of captive-bred parrots to release to the wild to increase the wild population [20]. Other adjustments have been made to the recovery programme; in particular, examining the fire regime at the breeding site of the parrot and which form of releasing captive birds into the wild was most effective (viz., spring release of adults to the breeding site, summer nest supplementation of eggs or nestlings, autumn fledgling release prior to winter migration, or winter release of adults/juveniles in wintering feeding areas). Controlled burns of the button-grass heath led to parrot food plants becoming more abundant. Also, the number of captive-bred fledglings released in autumn prior to northerly migration to overwintering on the mainland was increased [21].

Parrots returning to the summer breeding habitat have increased from 17 in 2017 to 70 in 2021 [22]. While the future is still uncertain, there is hope that the recovery programme is now working.

4 CONCLUDING COMMENTS

The aftermath reflects both the value and the limitations of environmental impact assessment of proposed developments. Project impact assessment was useful in the case of the proposed petrochemical plant: using constraint mapping, areas of environmental value were defined, avoidance of site development impacts could be achieved, protection of significant habitats could be incorporated in site planning, and compatibility with adjacent land uses could be managed.

In the assessment of the proposal to relocate the bulk chemical storage to Point Wilson, the potential for adverse impacts led to the selection of an alternative proposal. The inquiry into the East Coast Armaments Complex led to habitat fencing and increased culverts to facilitate greater tidal penetration to the upper saltmarsh.

However, project impact assessment was limited in what it could achieve in relation to the survival of the orange-bellied parrot. The contextual work for impact assessment defined the extent of the wintering habitat of the parrot, the local movement patterns of the parrot, and the ecohydrological connection to the catchment and seawater inundation of saltmarsh habitat. Also, baseline studies identified evidence of adverse impacts of earlier developments.

Notwithstanding, the threat to the orange-bellied parrot prime wintering site did stimulate action to address parrot recovery. This was not only the parrot recovery programme, but also the designation of wetlands related to habitat of the parrot and other bird species as a Ramsar site.

Parrot survival needed to be considered from a different perspective than project impact assessment. The parrot recovery programme considered the parrot ecosystem including the breeding areas in Southwest Tasmania, the migration paths to their wintering habitat, as well as the full range of its wintering feeding areas. There was a need for proactive interventions, such as captive breeding and reintroduction of captive-bred parrots back into the wild and addressing past effects of development such as fire regimes in parrot breeding areas and improving habitat areas through landholder agreements.

This type of proactive approach is beyond the scope of impact assessment. As stated in Martin et al. [20], there is the need for a decision process of: (i) intelligence – obtaining, processing, and delivering information to decision makers; (ii) promotion – recommending and mobilising support; (iii) prescription – turning policies into actions; (iv) invocation – implementing actions in a timely manner; (v) application – implementing the prescribed actions; (vi) appraisal – assessing whether prescriptions achieved the goal; and (vii) updating – revising prescriptions if required. They also identified a need for informed, empowered, and responsive governance and leadership. Martin and her colleagues indicate that these decision processes and governance arrangements were present for the Orange-Bellied Parrot Recovery Programme.

REFERENCES

[1] McHarg, I., *Design with Nature*, Natural History Press: New York, 1969.

[2] Chandler, C., Lyon, R. & Kinhill Planners, Avifauna study ICI Point Wilson development progress notes March to August 1980. Kinhill Planners: Melbourne, 1980.

[3] CI Australia, Point Wilson rezoning: Environmental effects statement. Kinhill Planners: Melbourne, 1978.

[4] Loyn, R. & Kinhill Planners, Historical records of orange-bellied parrots. Kinhill Planners: Melbourne, 1980.

[5] Carr, G. & Kinhill Planners, Survey of Victorian coastal saltmarsh in relation to the habitat of the orange-bellied parrot. Kinhill Planners: Melbourne, 1979.

[6] Loyn, R. & Kinhill Planners, Orange-bellied parrot count, August 12, 1979. Kinhill Planners: Melbourne, 1979.

[7] Lane, B. & Kinhill Planners, Results of a study of the orange-bellied parrot in south-west Tasmania. Kinhill Planners: Melbourne, 1979.

[8] Kinhill Planners, The hydrology of The Spit, Point Wilson. Kinhill Planners: Melbourne, 1982.

[9] Kinhill Planners, Point Wilson plan of development. ICI Australia: Melbourne, 1979.

[10] Maunsell & Partners, Point Lillias port and bulk liquid chemical storage facility: Environmental effects statement. Point Lillias Project Unit: Melbourne, 1995.

[11] Parliamentary Standing Committee on Public Works, East Coast Armaments Complex Point Wilson, Victoria. Commonwealth of Australia: Canberra, 1998.

[12] Barro Group, Environmental effects statement, Point Wilson quarry extension. Barro Group: Carlton, Victoria, 2008.

[13] Minister for Planning, Mountain View quarry extension assessment under the Environmental Effects Act 1978. Victorian Government: Melbourne, 2009.

[14] Orange-Bellied Parrot Recovery Team, Background and implementation information for the Orange-Bellied Parrot Recovery Plan. Department of Primary Industries and Water: Hobart, 2006.

[15] Casanelia, S., Information sheet on Ramsar Wetlands: Port Phillip Bay (western shoreline) and Bellarine Peninsula. Parks Victoria: Kew, 1999.

[16] Department of Environment, Land, Water and Planning, National recovery plan for the orange-bellied parrot *Neophema chrysogaster*. Australian Government: Canberra, 2016.

[17] Mathews, G., *The Birds of Australia*, **6**, p. 439, plate 316, 1917.

[18] Jarman, H., The orange-breasted parrot. *Australian Bird Watcher*, **2**, pp. 155–167, 1965.

[19] White, M., Menkhorst, P., Griffioen, P., Green, B., Salkin, D. & Pritchard, R., Orange-bellied parrot: A retrospective analysis of winter habitat availability. Arthur Rylah Institute for Environmental Research: Melbourne, 2016.

[20] Martin, T., Nally, S., Burbridge, A., Arnall, S., Garnett, S., Hayward, M., Lumsden, L., Menkhorst, P., McDonald-Maddern, E. & Possingham, H., Acting fast helps avoid extinction. *Conservation Letters*, **5**, pp. 274–280, 2012.

[21] Pritchard, R., Kelly, E., Biggs, J., Everaardt, A., Loyn, R., Magrath, M., Menkhorst, P., Hogg, C. & Geary, W., Identifying cost-effective recovery actions for a critically endangered species. *Conservation Science and Practice*, **4**, e546, 2022.

[22] Perkins, M., The parrot clawing its way back from the brink, one nest at a time. *The Age*, 10 Apr. 2022.

TRENDS IN THE DEVELOPMENT OF HYDROXYAPATITE FROM NATURAL SOURCES FOR BIOMEDICAL APPLICATIONS

DAVID O. OBADA[1,2,3*], ABUBAKAR ALIYU[2,3], UMAR M. KUTA[3,4], AYUBA LISIYAS[3,4], VANESSA F. OGENYI[3,4], MD. OSIM AQUATAR[6], SAMUEL JOSEPH[3,4], ISMAIL A. ADEWOLE[3,4], SABIU M. MUSA[3,4], JA'AFAR M. MA'AJI[3,5], CHRISTIANA P. SAMBO[3,4], AL-MUSTAFA M. SANI[3,5], HAMZA MUSA[3,5], FADILA A. BOLANLE[3,4], ISHAQ ADAM[3,4], TOCHUKWU C. ISAIAH[3,4], IDRIS B. ALHASSAN[3,4], RASHIDAT M. SALAWU[3,5], HAJARA ALIYU[3,4], AISHA ALIYU[3,4], MARYAM W. IBRAHIM[3,4], BILKISU AMINU[3,4], SARAH Y. SULE[1,3], AYODEJI N. OYEDEJI[1,3], KAZEEM A. SALAMI[1,3], ETUKESSIEN S. AKPAN[3,7], EMMANUEL T. SAMUEL[3] & LAMINU S. KUBURI[1,3]

[1]Department of Mechanical Engineering, Ahmadu Bello University, Nigeria
[2]Africa Centre of Excellence on New Pedagogies in Engineering Education, Ahmadu Bello University, Nigeria
[3]Multifunctional Materials Laboratory, Shell Office in Mechanical Engineering, Ahmadu Bello University, Nigeria
[4]Department of Polymer and Textile Engineering, Ahmadu Bello University, Nigeria
[5]Department of Metallurgical and Materials Engineering, Ahmadu Bello University, Nigeria
[6]Academy of Scientific and Innovative Research (AcSIR), Environmental Materials Division, CSIR-NEERI, India
[7]National Research Institute for Chemical Technology, Nigeria

ABSTRACT

Hydroxyapatite (HAp), a member of the apatite family is similar in structural and chemical composition to the natural bone. HAp is bioactive and biocompatible and has been used for several biomedical applications. Synthesizing HAp from natural sources can contribute significantly towards natural resource management, health care, waste utilization amongst others. Therefore, in this summarized review, we highlight the classification of HAp, discuss the sources and synthesis of hydroxyapatite, enumerate some applications of hydroxyapatite and touch on polymeric scaffolding materials based on naturally synthesized hydroxyapatite. The overall idea of this article is to summarize recent studies on this subject and link it to the waste management potentials of natural biowastes for the synthesis and biomedical applications of HAp.

Keywords: hydroxyapatite, biowastes, biocompatible, bioactive, scaffolds, polymers, waste valorization, natural sources, implants, apatite.

1 INTRODUCTION

Hydroxyapatite (HAp) with the chemical formula $Ca_{10}(PO_4)_6(OH)_2$, is one of the most versatile classes of biomaterials available today. Its chemical composition in terms of calcium–phosphate (Ca/P) ratio and other organic matter bears very close resemblance to that of the human bone [1]. This calcium phosphate salt is the chief inorganic mineral component of the human bone, enamel, and dentine [1], [2]. HAp is a calcium phosphate (CaP) which is the nomenclature of a family of minerals containing calcium cations (Ca^{2+}), orthophosphate (PO_4^{3-}), metaphosphate (PO^-_3), or pyrophosphate ($P_2O_7^{4-}$) anions, and sometimes hydrogen (H^+) or hydroxide (OH^-) ions. Consequently, hydroxyapatites are bioactive, biocompatible and bioresorbable materials that can be successfully adapted for biomedical applications. They have the characteristics to synergize with biological systems, with the potential to produce devices for assessment, treatment, supplementation, and/or substitution of ailing bodily tissues or organs. Hydroxyapatites have increasingly gained notable attention owing to their relevance in addressing the medical needs of ageing people (for instance, treating osteoporotic patients), and management of defective hard and soft tissues. HAp finds

* *ORCID: http://orcid.org/0000-0003-0886-0315*

www.witpress.com, ISSN 1743-3541 (on-line)
doi:10.2495/WMEI220171

extensive use in the areas of treating rickets diseases, autografts and allografts, promotion of rapid bone growth and union, drug delivery, etc. [3]. The trend in the development of hydroxyapatite from natural sources now saves the environment from solid waste pollution which may likely lead to land degradation [2], [4], and the effective utilization of these natural sources in HAp synthesis through environmentally friendly processes, now creates value addition from hitherto waste products [5]–[18]. HAp has been successfully devised as grafts for hard and soft tissues, and for drug transport systems. Furthermore, HAp acts to rejuvenate, damaged tissues through bio-integration with the site and location of the diseased part, which helps to lengthen the life expectancy of the patient [19]. The excellent resemblance in the biochemical configuration of HAp with that of the human bone minerals makes them very useful in orthopaedics and odontology, and their ability to maintain a thermodynamically stable crystalline state within their host tissue can be attributed to their interfacial stability and good adhesion potentials which rapidly leads to osseointegration [20]. Literature studies have reported and shown that HAp does not result in any inflammatory, toxic, or carcinogenic reactions to the host tissues surrounding it [1], [2], [5]–[18], [21], [22]. More so, it promotes osteoblast proliferation, osteointegration and forms a direct bond between an implant and the newly formed bone tissue when used in maxillofacial orthopaedic and dental applications [23]. Currently, hydroxyapatite is also being used to carry out drug delivery in human systems, as therapeutic agents etc. [24], [25]. From these, it has been established that HAp has enjoyed an extensive use in the biomedical fields, predominantly in dentistry, disorderliness and deformities of the bone and joints, and as implant coatings.

In the context of waste valorization and its environmental impact, the utilization of HAp synthesized from natural sources is environmentally sustainable, economical, and friendly, since natural biowastes are in abundance. For instance, the bones of a bovine specie in Nigeria which are typically made up of the skeleton, head, horns, hooves, and feet constitutes the second largest part by weight (approximately 36%) of bovine during processing at the abattoir [26]. Since the bones are rich in calcium and phosphorus, a large portion of the waste bones are used as poultry feed. The left-over waste bones can be further utilized for the synthesis of natural HAp for biomedical applications, and this approach could eventually outweigh the use of bovine derived waste bones in other sectors. This can contribute significantly towards an improved economy, cleaner environment and enhanced general health. Therefore, in this summarized review, we highlight the classification of HAp, discuss the sources and synthesis of hydroxyapatite, enumerate some applications of hydroxyapatite and touch on polymeric scaffolding materials based on naturally synthesized hydroxyapatite.

1.1 Classification of HAp sources

HAp can be derived from either natural or synthetic sources, forms a strong bond with host tissues making it quite useful for bone remodelling applications [1], [2], [6]. Natural sources include all biomaterials such as animal bones, fish bones and scales, shells, minerals, and plants. The synthetic sources are achieved by mixing reagents of calcium (Ca^{2+}) and phosphates (PO_4^{3}-) precursors, then allowing conditions for nucleation and crystal growth [20]. Fig. 1 outlines a summary of the sources of hydroxyapatites. Natural hydroxyapatite is typically sourced from the discards of animal bones such as bovine [27], fish bones [18], marine shells [28], eggshells [29], wood [30], mineral sources e.g., limestone [31] amongst many other natural sources. A lot of studies have synthesized HAp from these two sources (natural and synthetic) with different characteristics. Sadat-Shojai et al. [20] reported that HAp materials can be produced either chemically or derived from natural sources. Procedures for the syntheses of HAp with intended features should be pre-planned. Moreover, a lot of tailored syntheses protocol have been established overtime, yet the formulation of HAp materials with targeted expected characteristics has remained a puzzle due to the inclusion of undesirable intermediate or by-products during synthesis. Hence research on the synthesis and characterization of HAp for biomedical applications will remain a continuous activity. Relating these to the state-of-the art, synthetic and natural sources of HAp are being considered either alone or combined for bone tissue regeneration. Currently, dopants such as strontium (Sr), iron (Fe), copper (Cu), zeolites, etc. have been suggested to improve the osteogenesis and angiogenesis ability of hydroxyapatite-derived scaffolds [17]. Unlike synthetic HAp, the natural HAp is non-stoichiometric due to the presence of characteristic trace elements such as Na^+, Zn^{2+}, Mg^{2+}, K^+, Si^{2+}, Ba^{2+}, F^-, and CO_3^{2-} making it comparable to the biochemical configuration of the human bone [2], [17].Worthy of note is that HAp and its production techniques have very strong correlation with its properties, particularly its calcium to phosphate (Ca/P) ratio, elemental composition, surface and structural morphology, crystallinity and phases. These properties are also a function of its processing parameters such as temperature, holding time and sieve size [18]. These parameters all have their influences on the overall characteristics and application-based properties of the synthesized HAp.

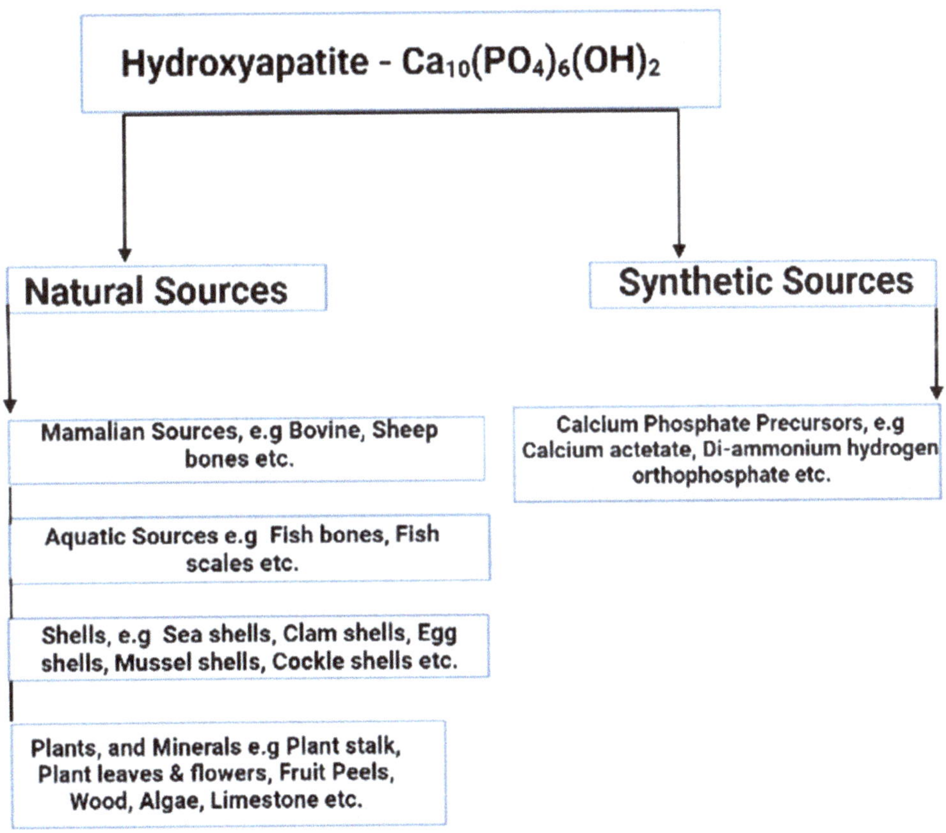

Figure 1: Sources of hydroxyapatite.

Typically, naturally synthesized HAp often normally deviates from the stoichiometric Ca/P ratio value of HAp being 1.67 [2]. This has been attributed to the inherent presence of trace elements in the composition which imitates the apatite found in the human bone. The incidence of these trace elements is however critical for biodegradation and bio-resorption during bioactive interactions, which then leads to total bone restoration in the patient. For example, there have been instances of improved activity per osteogenesis and bio-resorption in bone restoration using rare-earth metals doped HAp [32]. Reports and reviews from Obada et al. [17] and Pu'ad et al. [33] reported that bone regeneration and formation processes were accelerated by the inclusion of trace elements in HAp.

1.2 Extraction of HAp from natural sources

The extraction of HAp from natural sources particularly with interests in animal bone and catfish bone wastes was reported by our group [2], [5], [6], [11]–[18]. The usual step was with a pre-treatment approach of the bones by washing off dirt, fatty-proteinaceous matter, and cholesterol from the surfaces of the bones/precursors under running water. Next, the bones are boiled in hot water for about 8 h or more to ensure that embedded organic matter are further removed. Some literatures have further reported the addition of solvents like acetone, etc. in the boiling water for more effective removal, while others added surfactant and alkali solutions to detach the soft tissues and break their bonds from the bones [33]. The general extraction method of HAp from natural sources is the solid-state sintering approach. Others employ combining the calcination process with other methods such as sol-gel or wet chemical precipitation processes. Overall, it is desirable to ensure that the possible inherent infections or pathogens are eliminated within the process of developing the final HAp which buttresses the need for a high processing temperature in some cases [2], [33]. Obada et al. [2] have shown that solid-state synthesis of natural sources of HAp at temperatures of 900°C with low compaction pressures of 500 Pa, can produce bio-ceramics with prominent HAp phases, and enhanced mechanical properties [2], [11]. Therefore, a focus on extracting HAp from mammalian and aquatic sources will be further reviewed in this article.

As reported by Barakat et al. [27], alkaline hydrothermal hydrolysis treatment was utilized to extract pure phase HAp from bovine. The extracted nano-sized HAp showed nanoflakes shape of a Ca/P ratio of 1.86. Their work further revealed that 20–25% NaOH as a pre-treatment process before calcination resulted in the development of pure Hap. Londoño-Restrepo et al. [34] employed a combination of hydrothermal and calcination process to produce a dehydroxylated crystallite phase Hap of irregular shapes at 700°C, and semi spherical shapes at 800°C. Ofudje et al. [35] extracted Hap from pig bones using the calcination protocol. The calcination temperatures were 600°C, 800°C and 1000°C. The data presented showed that pure crystalline Hap was produced with rod-like morphologies, with a Ca/P of 1.88 the highest after calcining at 1000°C. A detailed review by Pu'ad et al. [33], showed that the alkaline heat treatment method was used to produce Hap having Ca/P ratio >1.67, and irregular shaped particle sizes of 20–100 μm. Moreover, when the same method was applied in addition to the calcination process for the extraction of Hap from pig bones, Hap of irregular shape size in a range of 70–180 nm at 800°C was produced. Similarly, direct calcination process of camel bones by Jaber et al. [36] at 1000°C gave an irregular shaped particle size range of 79–97 nm, with a calculated Ca/P ratio of 1.66 which is very close to the stoichiometric Ca/P ratio. Reports by Rahavi et al. [37] showed the properties of Hap when extracted from camel and horse bones at 700°C calcination temperature. The morphologies showed irregular shaped Hap sizes of 97 nm for camel bone derived Hap, and 28 nm for horse bone derived Hap. A Ca/P ratio of 2.036 was obtained for camel bone derived

Hap, while a Ca/P ratio of 2.131 was obtained for horse bone derived Hap. In a study conducted by Obada et al. [17], the solution combustion synthesis (SCS), assisted with thermal treatment at 900°C was used for the extraction of Hap from animal bones. This protocol favoured the formation of Hap with lower crystallinity, but with increased micro hardness. Fig. 2 describes the method used for the extraction of Hap from animal bones using SCS as reported by Obada et al. [17], and all the post processes and characterization thereof.

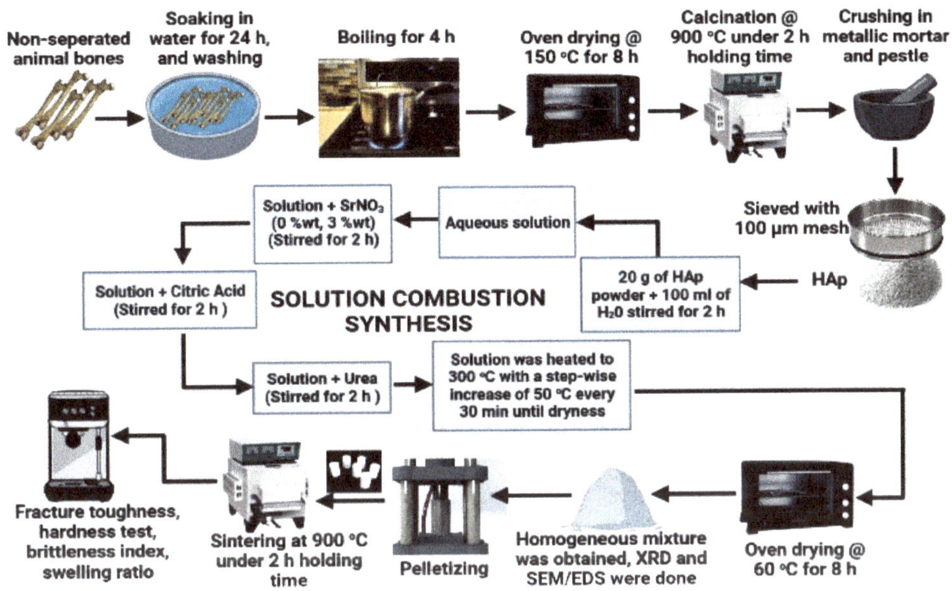

Figure 2: Schematic for the SCS protocol for the extraction and characterization of Hap [17].

As fish is consumed widely globally, there is a consequent increase in biowaste generation in the form of fish bones and scales. In our group, we have employed same pre-treatment method as used for animal bones to remove all unwanted debris from fish bones. Since fish bones and scales are rich in calcium, phosphate and carbonate, a lot of studies have engineered the extraction of Hap from these sources [13], [38]. Pal et al. [39] subjected *calcarifier* bones to temperatures from 200°C–1200°C, at an interval of 200°C for a holding time of 1 h. At higher sintering temperature of 1200°C, it was found that the Ca/P ratio of the sample was 1.62, a value lesser that the Ca/P ratio of stochiometric Hap by reason of the formation of tricalcium phosphate. Venkatesan et al. [40] extracted Hap from *Thunnus Obesus* fish using a combination of alkaline hydrolysis (AH) and calcination (CN) methods in the process. The tuna fish bone was subjected to heat treatment at 900°C for a holding time of 5 h. These methods (alkaline hydrolysis and calcination) were deemed suitable for producing Hap. Nonetheless, the Hap produced through the calcination process was more crystalline comparatively. Furthermore, Venkatesan et al. [40] reported that the AH method did not change the sizes of the particles. The Ca/P ratio of Hap derived via the calcination process was 1.65, a value quite close to that of stoichiometric Hap. Mondal et al. [41] subjected fish scales to the calcination process at 1000°C, followed by a 16 h milling procedure for the synthesis of Hap with Ca/P ratio of 1.71. The morphological features of the

Hap showed spherical shapes with an average size of 76.62 nm. We show the characterization result of the synthesis of Hap from *Clarias macrocephalus* fish bones. We used a handheld X-ray fluorescence (XRF) instrument from SPECTRO xSORT Combi (Kleve, Germany), with an energy dispersive SDD detector, an Rh-anode X-ray tube, a spot time of 3 mm, and an acquisition time of 60 s to elucidate the composition of the fish bones (see Table 1). The concentration of major and trace elements was determined. The accuracy of the hXRF instrument was ensured by performing several measurements. It can be observed from Table 1 that as reported by several researchers, extraction of Hap from natural fish bones contains trace elements like Mn, Fe, Sr. Zn, etc., with calcium and phosphorus the main elements in the composition. It has been reported that mammalian bones are composed of more trace elements and ions [33]. The concentration of these trace elements in a natural source varies by reason of differences in source, species, etc.

Table 1: XRF data for the composition of *Clarias macrocephalus* fish bones.

Element	Symbol	Concentration (%)
Silicon	Si	0.52
Phosphorus	P	26.4
Sulphur	S	0.06
Calcium	Ca	47.2
Manganese	Mn	0.02
Iron	Fe	0.02
Strontium	Sr	0.07
Zinc	Zn	0.02

We complement this data by reporting the morphological features (see Fig. 3) of the extracted Hap calcined at 900°C, using a scanning electron microscope (SEM machine model no-S2500, Hitachi, Japan) equipped with an electron dispersive spectrometer. The SEM images are complemented by EDS spectra as shown in Fig. 4. It can be observed that irregular and spherical shapes with agglomerated portions can be observed in the SEM images (Fig. 3(a) and (b)). This data is in line with summarized characteristics of Hap synthesized from natural sources as shown in Tables 2 and 3.

Figure 3: SEM images of *Clarias macrocephalus* fish bones. (a) Magnification of 20.00 kX; and (b) Magnification of 50.00 kX.

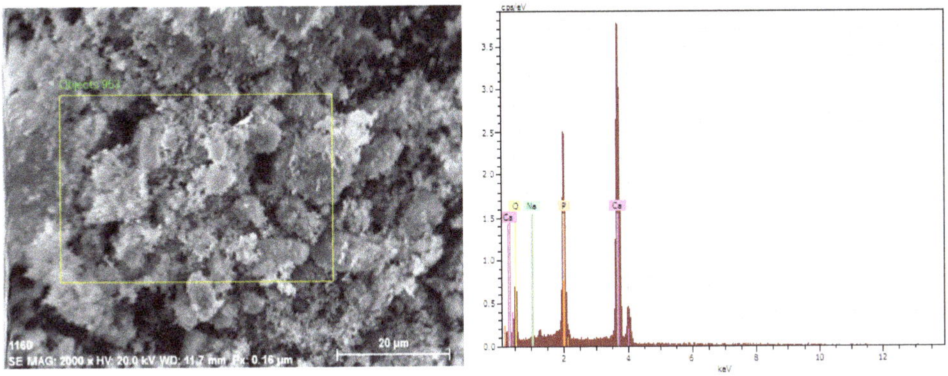

Figure 4: SEM/EDS spectra of *Clarias macrocephalus* fish bones.

Table 2: Characteristics of Hap extracted/synthesized by different protocols from natural sources (mammalian).

Methods			
Properties	Alkaline hydrolysis (AH)	Calcination ©	Combination (AH + C)
Sizes of the particles	20–100 µm	28–900 nm	70–700 nm
Morphological features	Irregular and flake-like in most cases	Irregular and rod like in most cases	Irregular, needle-like, and plate lie in most cases
Crystallinity	Low	High	High
Ca/P ratio	1.67–1.86	1.50–2.13	1.68–1.72
Trace elements	Yes (e.g. Sr, Mg, Na etc.)	Yes (e.g. Sr, Mg, Na, Al etc.)	Not reported

Table 3: Characteristics of HAp extracted/synthesized by different protocols from natural sources (fish bones and fish scales).

Methods			
Properties	Alkaline heat treatment (AHT)	Calcinati©(C)	Combination (AHT + C) **Only for fish scales*
Sizes of the particles	5–20 nm width; 17–100 nm length	5–76.2 nm	76.2 nm
Morphological features	Flat plate, rod like in most cases	Irregular, nearly spherical, agglomerate	Agglomerate
Crystallinity	Not reported	High [11]	Not reported
Ca/P ratio	1.76–2.01	1.62–1.71	1.62
Trace elements	Not reported	Mn, Fe, Sr, Na, Zn (this work)	Not reported

The EDS spectra of the *Clarias macrocephalus* fish bones corroborates the inclusion of trace elements in HAp extracted from aquatic natural sources and is in line with data obtained as shown in Table 1. The inclusion of sodium (Na) is noticed in the EDS spectra.

Tables 2 and 3 show summarized characteristics of HAp extracted/synthesized by different protocols from mammalian and aquatic sources.

2 APPLICABILITY OF HYDROXYAPATITE IN BIOMEDICINE AND POLYMERIC SCAFFOLDING MATERIaLS

HAp has stirred incredible interest in the field of biomedicine for various applications, noticeable among them is the drug delivery ability of HAp in providing antibiotics to hard tissues. Bi et al. [42] studied the fabrication and characterization of hydroxyapatite/sodium alginate/chitosan (HA/SA/CS) composite for drug delivery. The result of the study revealed that the structure of the developed composite improved the quality of the loaded drugs. Furthermore, HAp was used to coat metal inserts with a view to improving their biocompatibility. De Oliveira et al. [43] investigated the response of a nano-HAp coated implant via gene expression. The analyses they conducted showed that the use of nano-HAp coated implants can promote new bone formation in diabetic animals which is considered as a better alternative to the use of machined or double etched surfaces. Another industrial applicability of HAp is in the production of solid scaffolds for tissue engineering. Ergul et al. [44] used a 3D printer to fabricate hydrogels based on chitosan/poly (vinyl alcohol) with the inclusion of synthesized hydroxyapatite scaffolds for hard-tissue engineering. The study concluded that through the addition of HAp to scaffolds production, there is a promising innovation for bone tissue engineering.

In our group, we have attempted to use Polylactic acid (PLA) reinforced with HAp extracted from catfish bones to synthesize a biodegradable polymer that can be useful for biomedical applications. PLA is a biodegradable thermoplastic aliphatic polyester which is easily processed using a variety of protocols and has been used for an array of biomedical applications. However, because one of the key disadvantages of PLA is its low hydrophilicity, this results in a lack of cell attachment and interaction between the polymer and the surrounding tissues, as well as poor wetting qualities [45]. PLA bioactivity must be improved for it to be used in the intended bone regeneration application, considering all these features. The addition of natural HAp and trace elements like Sr [17] to PLA matrix can improve the 3D structures' hydrophilicity, osteoconductivity, mineralization after implantation, and mechanical characteristics. We performed these experiments using the melt extrusion technique. The melt extrusion/hot pressing manufacturing method is a method of turning raw materials, usually containing a polymer substrate and plasticizer, into a product that has uniform shapes and densities by pushing them through a heated barrel at a controlled temperature and pressure. Fig. 5 shows a description of the methods used in the development of the natural HAp-Sr reinforced polymer and all the characterization thereof. The structural data obtained for the composite using this process reveal the incorporation of the bioactive natural derived HAp with a semicrystalline characteristic (not shown).

3 CONCLUSION

This review summarizes trends in the synthesis of HAp from natural sources such as mammalian and aquatic amongst other sources like minerals, shells, plants and algae. We highlighted the differences in characteristics of natural HAp produced from different sources in terms of crystallinity, shapes, and sizes. The methods of extraction were crucial in tailoring the characteristics of the obtained HAp. The applications of HAp for biomedical applications

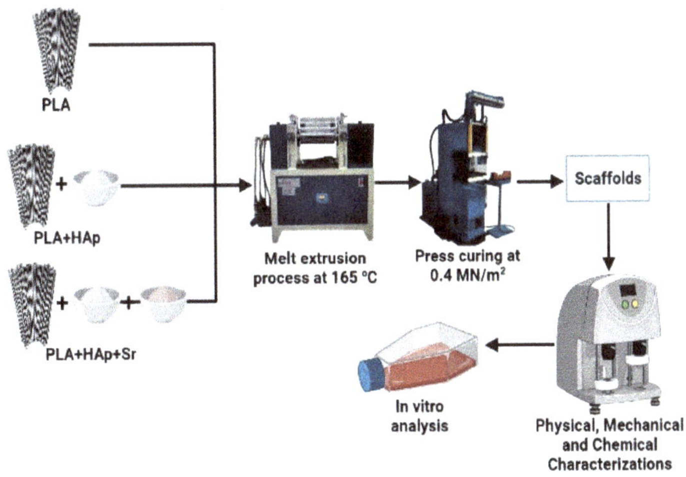

Figure 5: Methodological procedure to produce PLA-HAp-Sr scaffolds.

are ever increasing, thus a need to engineer more practical solutions like the development of natural HAp reinforced biodegradable polymers. It is therefore important to conclude that the extraction of HAp from natural sources ensures a cleaner environment and can be sustained because these natural sources can enhance the recovery of nutrients from biowastes to transform them into value added products.

ACKNOWLEDGEMENTS

The authors wish to acknowledge TETFund Nigeria for supporting this research under grant Ref: NRF_SETI_HSW_00714, 2020. Also, the authors would like to thank Hannah O. Olumoyegun for the administrative support given to all the Industrial Training Students (co-authors in this article) of 2021 at the Multifunctional Materials Laboratory, Shell Office Complex, Department of Mechanical Engineering, Ahmadu Bello University, Zaria, Nigeria.

REFERENCES

[1] Eliaz, N. & Metoki, N., Calcium phosphate bioceramics: A review of their history, structure, properties, coating technologies and biomedical applications. *Materials*, **10**(4), p. 334, 2017.

[2] Obada, D.O., Dauda, E.T., Abifarin, J.K., Dodoo-Arhin, D. & Bansod, N.D., Mechanical properties of natural hydroxyapatite using low cold compaction pressure: Effect of sintering temperature. *Materials Chemistry and Physics*, **239**, 122099, 2020.

[3] Bano, N., Jikan, S.S., Basri, H., Bakar, S.A.A. & Nuhu, A.H., Natural hydroxyapatite extracted from bovine bone. *Journal of Science and Technology*, **9**(2), 2017.

[4] Pandharipande, S.L. & Sondawale, M.S.S., Synthesis of hydroxyapatite from egg shell and preparation of bone like Bio-composites using it. *International Journal of Advanced Science and Technology*, **5**(8), pp. 36–47, 2016.

[5] Obada, D.O. et al., Fabrication of novel kaolin-reinforced hydroxyapatite scaffolds with robust compressive strengths for bone regeneration. *Applied Clay Science*, **215**, 106298, 2021.

[6] Obada, D.O., Dodoo-Arhin, D., Dauda, M., Anafi, F.O., Ahmed, A.S. & Ajayi, O.A., Physico-mechanical and gas permeability characteristics of kaolin based ceramic membranes prepared with a new pore-forming agent. *Applied Clay Science*, **150**, pp. 175–183, 2017.

[7] Horta, M.K.D.S., Moura, F.J., Aguilar, M.S., Westin, C.B., Navarro da Rocha, D. & de Campos, J.B., In vitro evaluation of natural hydroxyapatite from Osteoglossum bicirrhosum fish scales for biomedical application. *International Journal of Applied Ceramic Technology*, **18**(6), pp. 1930–1937, 2021.

[8] Deb, P., Barua, E., Lala, S.D. & Deoghare, A.B., Synthesis of hydroxyapatite from Labeo rohita fish scale for biomedical application. *Materials Today: Proceedings*, **15**, pp. 277–283, 2019.

[9] Mondal, S., Hoang, G., Manivasagan, P., Moorthy, M.S., Kim, H.H., Phan, T.T.V. & Oh, J., Comparative characterization of biogenic and chemical synthesized hydroxyapatite biomaterials for potential biomedical application. *Materials Chemistry and Physics*, **228**, pp. 344–356, 2019.

[10] Khamkongkaeo, A. et al., Sintering behavior and mechanical properties of hydroxyapatite ceramics prepared from Nile Tilapia (Oreochromis niloticus) bone and commercial powder for biomedical applications. *Ceramics International*, **47**(24), pp. 34575–34584, 2021.

[11] Obada, D.O., Dauda, E.T., Abifarin, J.K., Bansod, N.D. & Dodoo-Arhin, D., Mechanical measurements of pure and kaolin reinforced hydroxyapatite-derived scaffolds: a comparative study. *Materials Today: Proceedings*, **38**, pp. 2295–2300, 2021.

[12] Akpan, E.S., Dauda, M., Kuburi, L.S., Obada, D.O., Bansod, N.D. & Dodoo-Arhin, D., Hydroxyapatite ceramics prepared from two natural sources by direct thermal conversion: From material processing to mechanical measurements. *Materials Today: Proceedings*, **38**, pp. 2291–2294, 2021.

[13] Akpan, E.S., Dauda, M., Kuburi, L.S., Obada, D.O. & Dodoo-Arhin, D., A comparative study of the mechanical integrity of natural hydroxyapatite scaffolds prepared from two biogenic sources using a low compaction pressure method. *Results in Physics*, **17**, 103051, 2020.

[14] Abifarin, J.K., Obada, D.O., Dauda, E.T. & Dodoo-Arhin, D., Experimental data on the characterization of hydroxyapatite synthesized from biowastes. *Data in Brief*, **26**, 104485, 2019.

[15] Abifarin, J.K., Obada, D.O., Dauda, E.T. & Dodoo-Arhin, D., Experimental data on the characterization of hydroxyapatite synthesized from biowastes. *Data in Brief*, **26**, 104485, 2019.

[16] Osuchukwu, O.A., Salihi, A., Abdullahi, I. & Obada, D.O., Synthesis and characterization of sol-gel derived hydroxyapatite from a novel mix of two natural biowastes and their potentials for biomedical applications. *Materials Today: Proceedings*, 2022.

[17] Obada, D.O. et al., Solution combustion synthesis of strontium-doped hydroxyapatite: Effect of sintering and low compaction pressure on the mechanical properties and physiological stability. *Materials Letters*, **304**, 130613, 2021.

[18] Akpan, E.S., Dauda, M., Kuburi, L.S. & Obada, D.O., Box-Behnken experimental design for the process optimization of catfish bones derived hydroxyapatite: A pedagogical approach. *Materials Chemistry and Physics*, **272**, 124916, 2021.

[19] Billiet, T., Vandenhaute, M., Schelfhout, J., Van Vlierberghe, S. & Dubruel, P., A review of trends and limitations in hydrogel-rapid prototyping for tissue engineering. *Biomaterials*, **33**(26), pp. 6020–6041, 2012.

[20] Sadat-Shojai, M., Khorasani, M.T., Dinpanah-Khoshdargi, E. & Jamshidi, A., Synthesis methods for nanosized hydroxyapatite with diverse structures. *Acta Biomaterialia*, **9**(8), pp. 7591–7621, 2013.

[21] Okafor, E., Obada, D O., Ibrahim, Y. & Dodoo-Arhin, D., Prediction of the reflection intensity of natural hydroxyapatite using generalized linear model and ensemble learning methods. *Engineering Reports*, **3**(2), e12292, 2021.

[22] Okafor, E., Obada, D.O. & Dodoo-Arhin, D., Ensemble learning prediction of transmittance at different wavenumbers in natural hydroxyapatite. *Scientific African*, **9**, e00516, 2020.

[23] Fathi, M.H., Salehi, M., Mortazavi, V., Mousavi, S.B. & Parsapour, A., Novel hydroxyapatite/niobium surface coating for endodontic dental implant. *Surface Engineering*, **22**(5), pp. 353–358, 2006.

[24] Uskoković, V. & Uskoković, D.P., Nanosized hydroxyapatite and other calcium phosphates: Chemistry of formation and application as drug and gene delivery agents. *Journal of Biomedical Materials Research Part B: Applied Biomaterials*, **96**(1), pp. 152–191, 2011.

[25] Bertran, O. et al., Mineralization of DNA into nanoparticles of hydroxyapatite. *Dalton Transactions*, **43**(1), pp. 317–327, 2014.

[26] Omole, D.O. & Ogbiye, S.A., An evaluation of slaughterhouse wastes in south-west Nigeria. *American Journal of Environmental Protection*, **2**(3), pp. 85–89, 2013.

[27] Barakat, N.A., Khil, M.S., Omran, A.M., Sheikh, F.A. & Kim, H.Y., Extraction of pure natural hydroxyapatite from the bovine bones bio waste by three different methods. *Journal of Materials Processing Technology*, **209**(7), pp. 3408–3415, 2009.

[28] Shavandi, A., Bekhit, A.E.D.A., Ali, A. & Sun, Z., Synthesis of nano-hydroxyapatite (nHA) from waste mussel shells using a rapid microwave method. *Materials Chemistry and Physics*, **149**, pp. 607–616, 2015.

[29] Goloshchapov, D.L., Kashkarov, V.M., Rumyantseva, N.A., Seredin, P.V., Lenshin, A.S., Agapov, B.L. & Domashevskaya, E.P., Synthesis of nanocrystalline hydroxyapatite by precipitation using hen's eggshell. *Ceramics International*, **39**(4), pp. 4539–4549, 2013.

[30] Tampieri, A., Sprio, S., Ruffini, A., Celotti, G., Lesci, I.G. & Roveri, N., From wood to bone: Multi-step process to convert wood hierarchical structures into biomimetic hydroxyapatite scaffolds for bone tissue engineering. *Journal of Materials Chemistry*, **19**(28), pp. 4973–4980, 2009.

[31] Habibie, S., Wargadipura, A.H.S., Gustiono, D., Herdianto, N., Riswoko, A., Nikmatin, S. & Clarke, S., Production and characterization of hydroxyapatite bone substitute material performed from Indonesian limestone. *Int. J. Biomed. Eng. Sci*, **4**(1), pp. 11–23, 2017.

[32] Gao, J., Feng, L., Chen, B., Fu, B. & Zhu, M., The role of rare earth elements in bone tissue engineering scaffolds: A review. *Composites Part B: Engineering*, 109758, 2022.

[33] Pu'ad, N.M., Koshy, P., Abdullah, H.Z., Idris, M.I. & Lee, T.C., Syntheses of hydroxyapatite from natural sources. *Heliyon*, **5**(5), e01588, 2019.

[34] Londoño-Restrepo, S.M., Ramirez-Gutierrez, C.F., del Real, A., Rubio-Rosas, E. & Rodriguez-García, M.E., Study of bovine hydroxyapatite obtained by calcination at low heating rates and cooled in furnace air. *Journal of Materials Science*, **51**(9), pp. 4431–4441, 2016.

[35] Ofudje, E.A., Rajendran, A., Adeogun, A.I., Idowu, M.A., Kareem, S.O. & Pattanayak, D.K., Synthesis of organic derived hydroxyapatite scaffold from pig bone waste for tissue engineering applications. *Advanced Powder Technology*, **29**(1), pp. 1–8, 2018.

[36] Jaber, H.L., Hammood, A.S. & Parvin, N., Synthesis and characterization of hydroxyapatite powder from natural camelus bone. *Journal of the Australian Ceramic Society*, **54**(1), pp. 1–10, 2018.

[37] Rahavi, S.S., Ghaderi, O., Monshi, A. & Fathi, M.H., A comparative study on physicochemical properties of hydroxyapatite powders derived from natural and synthetic sources. *Russian Journal of Non-Ferrous Metals*, **58**(3), pp. 276–286, 2017.

[38] Akram, M., Ahmed, R., Shakir, I., Ibrahim, W.A.W. & Hussain, R., Extracting hydroxyapatite and its precursors from natural resources. *Journal of Materials Science*, **49**(4), pp. 1461–1475, 2014.

[39] Pal, A., Paul, S., Choudhury, A.R., Balla, V.K., Das, M. & Sinha, A., Synthesis of hydroxyapatite from lates calcarifer fish bone for biomedical applications. *Materials Letters*, **203**, pp. 89–92, 2017.

[40] Venkatesan, J., Qian, Z.J., Ryu, B., Thomas, N.V. & Kim, S.K., A comparative study of thermal calcination and an alkaline hydrolysis method in the isolation of hydroxyapatite from Thunnus obesus bone. *Biomedical Materials*, **6**(3), 035003, 2011.

[41] Mondal, B., Mondal, S., Mondal, A. & Mandal, N., Fish scale derived hydroxyapatite scaffold for bone tissue engineering. *Materials Characterization*, **121**, pp. 112–124, 2016.

[42] Bi, Y.G., Lin, Z.T. & Deng, S.T., Fabrication and characterization of hydroxyapatite/sodium alginate/chitosan composite microspheres for drug delivery and bone tissue engineering. *Materials Science and Engineering: C*, **100**, pp. 576–583, 2019.

[43] De Oliveira, P.G.F.P. et al., Influence of nano-hydroxyapatite coating implants on gene expression of osteogenic markers and micro-CT parameters: An in vivo study in diabetic rats. *Journal of Biomedical Materials Research Part A*, 2020.

[44] Ergul, N.M. et al., 3D printing of chitosan/poly (vinyl alcohol) hydrogel containing synthesized hydroxyapatite scaffolds for hard-tissue engineering. *Polymer Testing*, **79**, 106006, 2019.

[45] Shen, P., Moriya, A., Rajabzadeh, S., Maruyama, T. & Matsuyama, H., Improvement of the antifouling properties of poly (lactic acid) hollow fibre membranes with poly (lactic acid)–polyethylene glycol–poly (lactic acid) copolymers. *Desalination*, **325**, pp. 37–39, 2013.

Author index

www.ingramcontent.com/pod-product-compliance
Lightning Source LLC
LaVergne TN
LVHW062237200926
848484LV00006B/1559
9781784664633